18th International Conference on Recent Progress in Many-Body Theories 2015 (MBT18)

Journal of Physics: Conference Series Volume 702

Niagara Falls, New York, USA
16 – 21 August 2015

Editors:

Eckhard Krotscheck
Gerardo Ortiz

ISBN: 978-1-5108-2326-6
ISSN: 1742-6588

Printed from e-media with permission by:

Curran Associates, Inc.
57 Morehouse Lane
Red Hook, NY 12571

proceedings.com

Some format issues inherent in the e-media version may also appear in this print version.

Printed by Curran Associates, Inc. (2016)

For permission requests, please contact the Institute of Physics
at the address below.

Institute of Physics
Dirac House, Temple Back
Bristol BS1 6BE UK

Phone: 44 1 17 929 7481
Fax: 44 1 17 920 0979

techtracking@iop.org

Additional copies of this publication are available from:

Curran Associates, Inc.
57 Morehouse Lane
Red Hook, NY 12571 USA
Phone: 845-758-0400
Fax: 845-758-2633
Email: curran@proceedings.com
Web: www.proceedings.com

Table of contents

Volume 702

18th International Conference on Recent Progress in Many-Body Theories (MBT18)16–21 August 2015, Niagara Falls, USA

Accepted papers received: 14 March 2016Published online: Pl Online DATEPreface

Papers

Journal of Physics: Conference Series **702** (2016) 011001 doi:10.1088/1742-6596/702/1/011001

18th International Conference on Recent Progress in Many-Body Theories (MBT18)

E. Krotscheck[a] **and G. Ortiz**[b]

[a]Physics Department, University at Buffalo, The State University New York, Buffalo, NY 14260-1500

[b] Department of Physics, Indiana University, Bloomington, IN 47405-7105

E-mail: eckhardk@buffalo.edu

The 18th International Conference on Recent Progress in Many-Body Theories (MBT18) was held at the Conference Event Center, Niagara Falls, NY USA, in August 16-21, 2015. The conference continued the approximately biennial series initiated in Trieste in 1978 and was devoted to new developments in the field of many-body theories, including the emergent areas of quantum information and ultracold atom physics. The conference series encourages the exchange of ideas between physicists working in such diverse areas as nuclear and condensed matter physics, quantum chemistry, lattice Hamiltonians or quantum fluids. This volume includes the proceedings of the conference.

The MBT18 meeting was held back-to-back with the International Symposium on Quantum Fluids and Solids which was convened August 9-15, 2015 at the same location; between the two conferences a series of tutorial lectures intended for graduate students and young postdocs was organized. Lectures at these series were given by Frank Gasparini (Buffalo), Jordi Boronat (Barcelona), Erkki Thuneberg (Oulu) and Gerardo Ortiz (Indiana); the lecture notes are available at the conference website `http://www.physics.buffalo.edu/MBT18/tutorials.html` .

Many-body theories are an integral part in different fields of theoretical physics such as condensed matter, nuclear matter and field theory. Phase transitions and macroscopic quantum effects such as magnetism, Bose-Einstein condensation, superfluidity or superconductivity have been investigated within ultra-cold gases, finite systems or various nanomaterials. The conference series on Recent Progress in Many-Body Theories (RPMBT) is devoted to foster interaction and cross-fertilization between different fields, and also to discuss future lines of research.

The main conference program was split into eight sections:

- Many-Body Problems in Condensed Matter Physics
- Quantum Fluids and Solids
- Nuclear Many-Body and Relativistic Theories
- Cluster Physics and Intense Lasers
- Many-Body Aspects of Cold Gasses
- High Energy Density Matter

- New Developments in Computational Many-Body Techniques
- Quantum Phase Transitions and non-equilibrium Many-Body Theory

Almost ninety participants from nineteen countries participated. Sixty six contributions were presented, and thirty seven of them were invited talks. We were fortunate to have outstanding speakers which were able to convey the excitement of these fields, and would like to thank all them for making this conference so lively. As a particular highlight of the conference, the Eugene Feenberg Memorial Medal for outstanding results in the field of many-body theory, and the Hermann Kümmel Early Achievement Award in Many-Body Physics for young scientists in that field were awarded.

The Eugene Feenberg Medal was awarded to Christopher J. Pethick. The citation reads as follows: *For pioneering contributions and profound insights into many-body physics across diverse physical systems, ranging from ultracold atoms and quantum liquids to dense nuclear matter in neutron stars and stellar collapse.*

The Hermann Kümmel Early Achievement Award in Many-Body Physics was awarded to Lianyi He: *For pioneering work in the many-body theory of QCD at high isospin density and in the simulations of gauge fields in cold atoms.*

We were fortunate to have Christopher J. Pethick and Lianyi He present at the conference. We are particularly grateful to David Neilson and Joe Carlson for introducing Christopher Pethick and Lianyi He, respectively, and reviewing the history and outstanding accomplishments in their careers.

Further highlights of the conference were two special sessions to honor two esteemend members of our community, Gordon Baym and John W. Clark, at the occasion of their 80th Birthday.

The Conference continues the series of conferences held before in Trieste, Italy (1978); Oaxtapec, Mexico (1981); Odenthal-Altenberg, Germany (1983); San Francisco, USA (1985); Oulu, Finland (1987); Arad, Israel (1989); Minneapolis, USA (1991); Schlo Segau, Austria (1994); Sydney, Australia (1997); Seattle, USA (1999); Manchester, UK (2001); Santa Fe, USA (2004); Buenos Aires, Argentina (2005); Barcelona, Spain (2007); Columbus, USA (2009), Bariloche, Argentina (2011) and Rostock, Germany (2013).

We thank the International Advisory Committee, the International Programme Committee, and the Local Organizing Committee for their great support and advice. The next RPMBT meeting will take place in Korea in 2017.

The conference major funding, aside from registration fees, came through grants of the US National Science Foundation, and internal support from the Physics Department, the College of Arts and Sciences, and the office of the Vice President for Research of the University at Buffalo.

Eckhard Krotscheck and Gerardo Ortiz
Guest editors

Peer review statement

All papers published in this volume of *Journal of Physics: Conference Series* have been peer reviewed through processes administered by the proceedings Editors. Reviews were conducted by expert referees to the professional and scientific standards expected of a proceedings journal published by IOP Publishing.

Journal of Physics: Conference Series **702** (2016) 012001 doi:10.1088/1742-6596/702/1/012001

Frustrated Heisenberg antiferromagnet on the honeycomb lattice with spin quantum number $s \geq 1$

P H Y Li[1], **R F Bishop**[1] **and C E Campbell**[2]

[1] School of Physics and Astronomy, Schuster Building, The University of Manchester, Manchester, M13 9PL, UK

[2] School of Physics and Astronomy, University of Minnesota, 116 Church Street SE, Minneapolis, Minnesota 55455, USA

E-mail: raymond.bishop@manchester.ac.uk; peggyhyli@gmail.com

Abstract. The ground-state (GS) phase diagram of the frustrated spin-s J_1–J_2–J_3 Heisenberg antiferromagnet on the honeycomb lattice is studied using the coupled cluster method implemented to high orders of approximation, for spin quantum numbers $s = 1, \frac{3}{2}, 2, \frac{5}{2}$. The model has antiferromagnetic (AFM) nearest-neighbour, next-nearest-neighbour and next-next-nearest-neighbour exchange couplings (with strength $J_1 > 0$, $J_2 > 0$ and $J_3 > 0$, respectively). We specifically study the case $J_3 = J_2 = \kappa J_1$, in the range $0 \leq \kappa \leq 1$ of the frustration parameter, which includes the point of maximum classical ($s \to \infty$) frustration, viz., the classical critical point at $\kappa_{\rm cl} = \frac{1}{2}$, which separates the Néel phase for $\kappa < \kappa_{\rm cl}$ and the collinear striped AFM phase for $\kappa > \kappa_{\rm cl}$. Results are presented for the GS energy, magnetic order parameter and plaquette valence-bond crystal (PVBC) susceptibility. For all spins $s \geq \frac{3}{2}$ we find a quantum phase diagram very similar to the classical one, with a direct first-order transition between the two collinear AFM states at a value $\kappa_c(s)$ which is slightly greater than $\kappa_{\rm cl}$ [e.g., $\kappa_c(\frac{3}{2}) \approx 0.53(1)$] and which approaches it monotonically as $s \to \infty$. By contrast, for the case $s = 1$ the transition is split into two such that the stable GS phases are one with Néel AFM order for $\kappa < \kappa_{c_1} = 0.485(5)$ and one with striped AFM order for $\kappa > \kappa_{c_2} = 0.528(5)$, just as in the case $s = \frac{1}{2}$ (for which $\kappa_{c_1} \approx 0.47$ and $\kappa_{c_2} \approx 0.60$). For both the $s = \frac{1}{2}$ and $s = 1$ models the transition at κ_{c_2} appears to be of first-order type, while that at κ_{c_1} appears to be continuous. However, whereas in the $s = \frac{1}{2}$ case the intermediate phase appears to have PVBC order over the entire range $\kappa_{c_1} < \kappa < \kappa_{c_2}$, in the $s = 1$ case PVBC ordering either exists only over a very small part of the region or, more likely, is absent everywhere.

1. Introduction

Quantum spin-lattice models, in which the sites of a give regular periodic lattice are all occupied by magnetic ions with spin quantum number s, offer a rich arena for the study of exotic ground-state (GS) phases that are not present in their classical ($s \to \infty$) counterparts. Whereas interactions between the classical spins give rise to magnetic ground states in which the spins are ordered such that each individual spin is oriented in a specific direction, quantum fluctuations can act either to diminish the corresponding magnetic order parameter (viz., the average local onsite magnetization) or to destroy it altogether. In the former case, where the long-range order (LRO) is only partially reduced, such quasiclassical magnetically ordered states spontaneously break both the SU(2) spin-rotation and time-reversal symmetries.

By contrast, such intrinsically quantum-mechanical states as the various valence-bond crystalline (VBC) phases, in which specific combinations of the lattice spins combine into spin singlets, have zero magnetic order and break neither of the SU(2) spin-rotation and time-reversal symmetries, although they still break some lattice symmetries. Yet other states exist in which, for example, time-reversal symmetry is preserved, so that magnetic order is definitely absent, but for which the SU(2) spin-rotation symmetry is still broken. These are the so-called multipolar or spin-nematic phases. Finally, of course, one also has the possibility of quantum spin-liquid (QSL) phases that preserve all of the symmetries, including the lattice symmetries.

For a given regular lattice in d spatial dimensions, one is usually interested in the interplay between frustration (which may be tuned, for example, by varying the relative strengths of competing interactions in the model Hamiltonian that separately tend to promote different forms of magnetic LRO) and quantum fluctuations. In broad terms quantum fluctuations are larger for lower values of both the spatial dimensionality d and the spin quantum number s. They are typically also larger, for given spatial dimensionality d, for lattices with smaller values of the coordination number z.

For the case of $d = 1$ isotropic Heisenberg chain systems the Mermin-Wagner theorem [1] excludes the possibility of GS magnetic order even at zero temperature ($T = 0$), since it is impossible to break a continuous symmetry for any such system. The Mermin-Wagner theorem similarly implies the absence of magnetic LRO in any $d = 2$ isotropic Heisenberg system at all nonzero temperatures ($T > 0$). The behaviour and GS quantum phase structure of two-dimensional (2D) spin-lattice models at $T = 0$ has thus come to occupy a special role in the study of quantum phase transitions.

Spin-lattice systems are said to be frustrated when constraints are present that preclude the formation of a GS phase which satisfies all of the (generally, pairwise) interactions among the spins. Frustration is strongly associated with macroscopic degeneracy of the GS phase, with the consequent existence of strong quantum fluctuations among the states in the degenerate manifold. Either quantum or thermal fluctuations can then, in such a situation, suppress magnetic LRO, and the possibility of such exotic non-classical states as those discussed above forming the stable GS phase under certain conditions is heightened. The low coordination number, $z = 3$, of the honeycomb lattice further enhances the quantum fluctuations, and thereby makes it a special 2D spin-lattice model candidate for the study of its $T = 0$ GS quantum phase diagram when dynamical frustration is introduced via competing interactions.

In the present paper we study the so-called J_1–J_2–J_3 model on the honeycomb lattice with antiferromagnetic (AFM) Heisenberg exchange interactions between pairs of nearest-neighbour (NN) spins (of strength J_1), next-nearest-neighbour (NNN) spins (of strength J_2), and next-next-nearest-neighbour (NNNN) spins (of strength J_3). Even in the classical ($s \to \infty$) limit the model has a rich phase diagram [2, 3], as we discuss further in Sec. 2 below. In the case where all three bonds are AFM in nature (i.e., $J_i > 0$; $i = 1, 2, 3$), the classical system exhibits two collinear AFM phases, namely the so-called Néel and striped phases, as well as a spiral phase. The three phases meet in a triple point at $J_3 = J_2 = J_1/2$ (see, e.g., Refs. [3,4]). This is the point of maximum classical frustration, where the classical GS phase has macroscopic degeneracy. For the present study we consider the model along the line $J_3 = J_2 \equiv \kappa J_1$ for the case $J_1 > 0$, as a function of the frustration parameter κ in the range $0 \leq \kappa \leq 1$. The classical ($s \to \infty$) version of the model thus has a single quantum phase transition in its $T = 0$ phase diagram at $\kappa_{\rm cl} = \frac{1}{2}$. For $\kappa < \kappa_{\rm cl}$ the stable GS phase is the Néel AFM phase, whereas for $\kappa > \kappa_{\rm cl}$ the stable GS phase is the striped AFM phase. In fact, at $T = 0$, there actually exists an infinite family of non-coplanar states, all of which are degenerate in energy with the striped state. However, it is asserted [3] that both thermal and quantum fluctuations break this degeneracy in favour of the collinear striped state, at least in the large-s limit [2].

Whereas the spin-$\frac{1}{2}$ J_1–J_2–J_3 model on the honeycomb lattice, or particular cases of it (e.g., when $J_3 = J_2$ or $J_3 = 0$), have been investigated by many authors with a variety of theoretical tools (see, e.g., Refs. [2–24]), there are far fewer studies of the model in the case $s > \frac{1}{2}$. A particular exception is a very recent study [25] of the $s = 1$ J_1–J_2 model (i.e., when $J_3 = 0$) on the honeycomb lattice, using the density-matrix renormalization group (DMRG) method. Our specific aim here is to extend earlier work using the coupled cluster method (CCM) applied to the spin-$\frac{1}{2}$ version of the J_1–J_2–J_3 model on the honeycomb lattice along the line $J_3 = J_2$ [9,15,17], to cases $s > \frac{1}{2}$. In particular, we now compare results for the case $s = \frac{1}{2}$ with those for $s = 1, \frac{3}{2}, 2, \frac{5}{2}$.

We note that by now there exist many experimental realizations of frustrated honeycomb-lattice systems with AFM interactions. These include such magnetic compounds as $Na_3Cu_2SbO_6$ [26], $InCu_{2/3}V_{1/3}O_3$ [27], β-$Cu_2V_2O_7$ [28], and Cu_5SbO_6 [29], in each of which the $s = \frac{1}{2}$ Cu^{2+} ions are situated on the sites of weakly coupled honeycomb-lattice layers. The iridates A_2IrO_3 (A = Na, Li) [30–33] are also believed to be magnetically ordered Mott insulators in which the Ir^{4+} ions form effective $s = \frac{1}{2}$ moments arrayed on weakly-coupled honeycomb-lattice layers. Other similar honeycomb materials include, for example, the families of compounds $BaM_2(XO_4)_2$ (M = Co, Ni; X = P, As) [34] and $Cu_3M_2SbO_6$ (M = Co, Ni) [35]. In both of these families the magnetic ions M^{2+} are again disposed in weakly-coupled layers where they occupy the sites of a honeycomb lattice. In both families the Ni^{2+} ions appear to take the high-spin value $s = 1$, whereas the Co^{2+} ions appear to take the low-spin value $s = \frac{1}{2}$ in the former family $BaCo_2(XO_4)_2$ and the high-spin value $s = \frac{3}{2}$ in the latter compound $Cu_3Co_2SbO_6$. As a last example of an $s = \frac{3}{2}$ honeycomb-lattice AFM material, we also mention the layered compound $Bi_3Mn_4O_{12}(NO_3)$ [36,37] in which the spin-$\frac{3}{2}$ Mn^{4+} ions sit on the sites of the honeycomb layers.

The remainder of the paper is organized as follows. In Sec. 2 we discuss further the model itself, before we give a brief description in Sec. 3 of the CCM formalism that we apply to it. The results are then presented in Sec. 4, and we conclude in Sec. 5 with a discussion and summary.

2. The model

The Hamiltonian of the J_1–J_2–J_3 model on the honeycomb lattice is given by

$$H = J_1 \sum_{\langle i,j \rangle} \mathbf{s}_i \cdot \mathbf{s}_j + J_2 \sum_{\langle\langle i,k \rangle\rangle} \mathbf{s}_i \cdot \mathbf{s}_k + J_3 \sum_{\langle\langle\langle i,l \rangle\rangle\rangle} \mathbf{s}_i \cdot \mathbf{s}_l \,, \tag{1}$$

where the sums over $\langle i,j \rangle$, $\langle\langle i,k \rangle\rangle$ and $\langle\langle\langle i,l \rangle\rangle\rangle$ run over all NN, NNN and NNNN bonds, respectively, on the lattice, counting each pair of spins once and once only in each of the three sums. Each site i of the honeycomb lattice carries a spin-s particle described by the SU(2) spin operator $\mathbf{s}_i \equiv (s_i^x, s_i^y, s_i^z)$, with $\mathbf{s}_i^2 = s(s+1)$, and, for the cases considered here, $s = 1, \frac{3}{2}, 2, \frac{5}{2}$. The lattice and the Heisenberg exchange bonds are illustrated in Fig. 1(a). For the present study we are interested in the case where each of the three types of bonds is AFM in nature (i.e., $J_m > 0$; $m = 1,2,3$). Without loss of generality we may put $J_1 \equiv 1$ to set the overall energy scale, and we will specifically consider the case where $J_3 = J_2 \equiv \kappa J_1$, in the interesting window $0 \leq \kappa \leq 1$ of the frustration parameter κ.

The honeycomb lattice is bipartite, but comprises two triangular Bravais sublattices $\mathcal{A}$ and $\mathcal{B}$. The basis vectors $\mathbf{a} = \sqrt{3}d\hat{x}$ and $\mathbf{b} = (-\sqrt{3}\hat{x} + 3\hat{z})d/2$ are illustrated in Fig. 1(b), where the lattice is defined to lie in the xz plane as shown, and where d is the lattice spacing (i.e., the distance between NN sites). The unit cell i at position vector $\mathbf{R}_i = m_i\mathbf{a} + n_i\mathbf{b}$, where $m_i, n_i \in \mathbb{Z}$, now comprises the two sites at $\mathbf{R}_i \in \mathcal{A}$ and $(\mathbf{R}_i + d\hat{z}) \in \mathcal{B}$. The reciprocal lattice vectors corresponding to the real-space vectors $\mathbf{a}$ and $\mathbf{b}$ are thus $\boldsymbol{\alpha} = 2\pi(\sqrt{3}\hat{x} + \hat{z})/(3d)$ and $\boldsymbol{\beta} = 4\pi/(3d)\hat{z}$. The Wigner-Seitz unit cell and the first Brillouin zone are thus the parallelograms formed, respectively, by the pairs of vectors $(\mathbf{a}, \mathbf{b})$ and $(\boldsymbol{\alpha}, \boldsymbol{\beta})$. Both may equivalently be taken as being centred on a point of sixfold rotational symmetry in their respective spaces. Thus, the Wigner-Seitz unit cell may be taken as being bounded by the sides of a primitive hexagon of

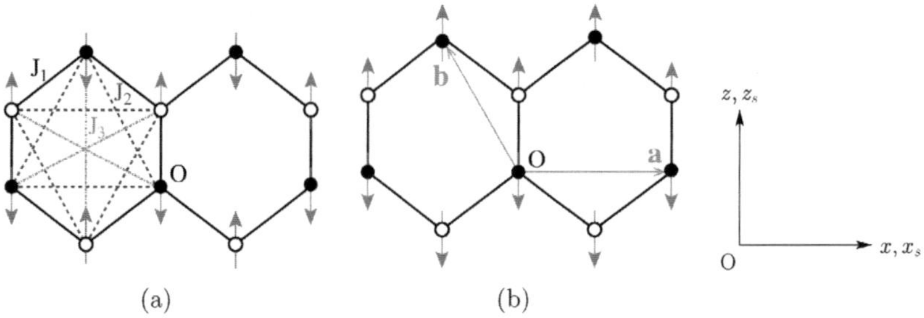

Figure 1. The J_1–J_2–J_3 model on the honeycomb lattice, showing (a) the bonds (J_1 = —, J_2 = - - -, J_3 = - · · -) and the Néel state, and (b) the triangular Bravais lattice vectors **a** and **b** and one of three equivalent striped states. Sites on the two triangular sublattices $\mathcal{A}$ and $\mathcal{B}$ are shown by filled and empty circles respectively, and the spins are represented by the (red) arrows on the lattice sites.

side length d as in Fig. 1. In this case the first Brillouin zone is also a hexagon, now of side length $4\pi/(3\sqrt{3}d)$, and which is also rotated by 90° with respect to the Wigner-Seitz hexagon.

The classical ($s \to \infty$) version of the J_1–J_2–J_3 model of Eq. (1) on the honeycomb lattice already itself displays a rich $T = 0$ GS phase diagram (see, e.g., Refs. [2,3,6]). The generic stable GS phase is a coplanar spiral configuration of spins defined by a wave vector $\mathbf{Q}$, together with an angle θ that is the relative orientation of the two spins in the same unit cell i characterized by the lattice vector $\mathbf{R}_i$. The two classical spins in unit cell i are given by

$$\mathbf{s}_{i,\rho} = -s[\cos(\mathbf{Q}\cdot\mathbf{R}_i + \theta_\rho)\hat{z}_s + \sin(\mathbf{Q}\cdot\mathbf{R}_i + \theta_\rho)\hat{x}_s]\,; \quad \rho = \mathcal{A},\, \mathcal{B}\,, \tag{2}$$

where $\hat{x}_s$ and $\hat{z}_s$ are two orthogonal unit vectors that define the spin-space plane, as shown in Fig. 1. We choose the two angles θ_ρ such that $\theta_\mathcal{A} = 0$ and $\theta_\mathcal{B} = \theta$.

When all three bonds are AFM in nature (i.e., $J_m > 0$; $m = 1,2,3$), as considered here, it has been shown [3,6] that the classical model has a $T = 0$ GS phase diagram consisting of three different phases. With reference to an origin at the centre of the hexagonal Wigner-Seitz cell, one may show that one value of the spiral wave vector $\mathbf{Q}$ that minimizes the classical GS energy of the model is given by

$$\mathbf{Q} = \frac{2}{\sqrt{3}d}\cos^{-1}\left[\frac{(J_1 - 2J_2)}{4(J_2 - J_3)}\right]\hat{x}\,, \tag{3}$$

together with $\theta = \pi$. Equation (3) is clearly only valid when

$$-1 \leq \frac{J_1 - 2J_2}{4(J_2 - J_3)} \leq 1\,. \tag{4}$$

If we define $x \equiv J_2/J_1$ and $y \equiv J_3/J_1$, Eq. (4) is equivalent to the inequalities,

$$y \leq \frac{3}{2}x - \frac{1}{4}\,; \quad y \leq \frac{1}{2}x + \frac{1}{4}\,. \tag{5}$$

In the positive quadrant (i.e., $x \geq 0$, $y \geq 0$) of the xy plane the classical model has the spiral phase described by the wave vector $\mathbf{Q}$ of Eq. (3) and $\theta = \pi$ as the stable GS phase in the region defined by Eq. (5).

Everywhere on the boundary line $y = \frac{3}{2}x - \frac{1}{4}$ of the spiral phase, $\mathbf{Q} = \mathbf{\Gamma} = (0,0)$, which simply describes the Néel AFM phase shown in Fig. 1(a). Similarly, everywhere on the other boundary line $y = \frac{1}{2}x + \frac{1}{4}$, $\mathbf{Q} = 2\pi/(\sqrt{3}d)\hat{x}$, which describes the collinear striped AFM phase shown in Fig. 1(b). Both the phase transitions between the spiral and Néel phases and between the spiral and striped phases are clearly continuous ones. The two phase boundaries meet at the tricritical point $(x, y) = (\frac{1}{2}, \frac{1}{2})$. Finally, one can also easily show that there is a first-order phase transition between the two collinear AFM phases along the line $x = \frac{1}{2}$, $y > \frac{1}{2}$. To summarize, in the regime where $J_1 > 0$ and $x \geq 0$, $y \geq 0$, the classical version of the honeycomb-lattice J_1–J_2–J_3 model has three stable GS phases at $T = 0$. These are: (a) a spiral phase for $0 < y < \frac{3}{2}x - \frac{1}{4}$, $\frac{1}{6} < x < \frac{1}{2}$ and $0 < y < \frac{1}{2}x + \frac{1}{4}$, $x > \frac{1}{2}$; (b) a Néel AFM phase for $y > 0$, $0 < x < \frac{1}{6}$ and $y > \frac{3}{2}x - \frac{1}{4}$, $\frac{1}{6} < x < \frac{1}{2}$; and (c) a striped collinear AFM phase for $y > \frac{1}{2}x + \frac{1}{4}$, $x > \frac{1}{2}$. Clearly, along the line $y = x$ considered here, which includes the tricritical point at $(\frac{1}{2}, \frac{1}{2})$, there are just two stable GS phases, namely the collinear Néel and striped AFM phases.

It is worth noting that both the spiral and the striped states described by the wave vector of Eq. (3) and its appropriate limiting form (and $\theta = \pi$) have two other equivalent states rotated by $\pm\frac{2}{3}\pi$ in the honeycomb xz plane. We also note that in the limiting case when the J_2 bond dominates (i.e., when $x \to \infty$ for a fixed finite value of y), the spiral pitch angle $\phi = \cos^{-1}[\frac{1}{4}(J_1 - 2J_2)/(J_2 - J_3)] \to \frac{2}{3}\pi$. Clearly, in this limit, the classical model reduces to two disconnected Heisenberg antiferromagnets (HAFs) on interpenetrating triangular lattices, each with the 3-sublattice Néel ordering of NN spins (on each triangular lattice) oriented at an angle $\frac{2}{3}\pi$ to one another. Precisely in this limit the wave vector $\mathbf{Q}$ of Eq. (3) becomes one of the six corners, $\mathbf{K}^{(1)} = 4\pi/(3\sqrt{3}d)\hat{x}$, of the hexagonal first Brillouin zone. The two inequivalent corner vectors thus describe the two distinct 3-sublattice Néel orderings for a classical triangular-lattice HAF.

For spiral pitch angles in the range $\frac{2}{3}\pi < \phi \leq \pi$ the wave vector $\mathbf{Q}$ of Eq. (3) lies outside the first hexagonal Brillouin zone. It can equivalently be mapped back inside this range of values, when $\mathbf{Q}$ then moves continuously from a corner at position $\mathbf{K}^{(3)} = 2\pi(-\hat{x} + \sqrt{3}\hat{z})/(3\sqrt{3}d)$ along an edge to its midpoint at $\mathbf{M}^{(2)} = 2\pi/(3d)\hat{z}$. Thus, the striped AFM state shown in Fig. 1(b) may equivalently be described by the ordering wave vector $\mathbf{Q} = \mathbf{M}^{(2)}$ (with the relative angle between the two triangular sublattices $\mathcal{A}$ and $\mathcal{B}$ being $\theta = \pi$). The two other equivalent striped states have wave vectors corresponding to the other two inequivalent midpoints of the hexagonal Brillouin zone edges, at $\mathbf{M}^{(1)} = \pi(\sqrt{3}\hat{x} + \hat{z})/(3d)$ and $\mathbf{M}^{(3)} = \pi(-\sqrt{3}\hat{x} + \hat{z})/(3d)$ (with $\theta = 0$ in these two cases).

While Eq. (2) is generally sufficient to describe the classical GS spin configuration [38], it relies on the assumption that the GS order either is unique (up to a global rotation of all spins by the same amount) or exhibits, at most, a discrete degeneracy (e.g., as associated with rotations of the wave vector $\mathbf{Q}$ by $\pm 120°$ about the $\hat{y}$ axis). Nevertheless, the assumption can be shown to be false for special values of $\mathbf{Q}$ [3, 38], which include the cases when $\mathbf{Q}$ equals either one half or one quarter of a reciprocal lattice vector $\mathbf{G}$. This includes precisely the case for the striped states for which the wave vectors $\mathbf{Q} = \mathbf{M}^{(i)}$, $i = 1, 2, 3$, equal one half of corresponding reciprocal lattice vectors. In this case, it has been shown [3] that the GS ordering now spans a 2D manifold of non-planar spin configurations, all of which are degenerate in energy with those of the striped states.

Classical spin-lattice systems that display such an infinitely degenerate family (IDF) of GS phases in some region of their $T = 0$ phase space are well known to be prime candidates for the emergence of novel quantum phases with no classical counterparts in the corresponding quantum systems. Quantum fluctuations then often act to lift this accidental GS degeneracy (either wholly or in part) by the order by disorder mechanism [38–40] in favour of just one (or several) member(s) of the classical IDF. As we noted previously in Sec. 1, the striped collinear state is indeed energetically selected by quantum fluctuations [3] in the present case in the large-s limit [2] where first-order linear spin-wave theory (LSWT) becomes exact.

Of course, quantum fluctuations can also be expected in such cases of macroscopic degeneracy of the classical GS phase, to destroy completely the magnetic LRO, as we discussed in Sec. 1, Clearly, this is most likely to occur for small values of s, when the results of LSWT become less likely to remain valid and when quantum fluctuations become larger. Since the specific case when $J_3 = J_2 \equiv \kappa J_1$ includes the classical tricritical point at $\kappa_{\rm cl} = \frac{1}{2}$, which is the point of maximum classical frustration, we restrict further attention to this potentially rich regime in the entire parameter space of the J_1–J_2–J_3 model. Along this line, the classical ($s \to \infty$) model at $T = 0$ undergoes a first-order transition from the Néel phase, which is the stable GS phase for $\kappa < \kappa_{\rm cl}$, to the striped phase which is the stable GS phase for $\kappa > \kappa_{\rm cl}$. From our above discussion it is clear that the most promising regime for novel quantum states, with no classical counterparts, to emerge is the region around $\kappa \approx \frac{1}{2}$.

In an earlier paper [9] the J_1–J_2–J_3 model with $J_3 = J_2 \equiv \kappa J_1 > 0$ was studied for the case $s = \frac{1}{2}$ in the window $0 \leq \kappa \leq 1$, using the CCM. It was found [9] that the classical ($s \to \infty$) transition at $\kappa_{\rm cl} = \frac{1}{2}$ is split in the $s = \frac{1}{2}$ case into two transitions at $\kappa_{c_1} < \kappa_{\rm cl}$ and $\kappa_{c_2} > \kappa_{\rm cl}$, with the Néel phase surviving for $\kappa < \kappa_{c_1} \approx 0.47$ and the striped phase for $\kappa > \kappa_{c_2} \approx 0.60$. A paramagnetic phase, with no discernible magnetic LRO, was indeed found to exist in the intermediate regime $\kappa_{c_1} < \kappa < \kappa_{c_2}$. CCM calculations were also performed to measure the susceptibilities of the two AFM phases on either side of the paramagnetic regime against the formation of plaquette valence-bond crystalline (PVBC) order [9]. It was thereby concluded that the paramagnetic state was most likely one with PVBC order over the entire intermediate regime $\kappa_{c_1} < \kappa < \kappa_{c_2}$. On the basis of all the CCM calculations (i.e., for the GS energy per spin E/N, the GS magnetic order parameter M, and the susceptibility χ_p against the formation of PVBC order, for the two AFM states on either side of the intermediate regime), the accumulated evidence pointed towards the quantum phase transition (QPT) at the quantum critical point (QCP) $\kappa = \kappa_{c_2}$ being a first-order one, just as is the classical phase transition at $\kappa = \kappa_{\rm cl}$. By contrast, the QPT at $\kappa = \kappa_{c_1}$ appeared to be a continuous one on the basis of the CCM results presented. Since the quasiclassical Néel phase and the quantum PVBC phase break different symmetries, however, the usual Landau-Ginzburg-Wilson scenario of continuous phase transitions is inapplicable, and it was suggested that the CCM results [9] provide strong evidence for the QPT at $\kappa = \kappa_{c_1}$ being of the deconfined quantum critical type [41, 42].

In view of the qualitative differences between the GS phase diagrams of the above spin-$\frac{1}{2}$ and classical ($s \to \infty$) versions of the model, it is obviously of considerable interest to examine the model in the case where the spin quantum number $s > \frac{1}{2}$. One of the great strengths of the CCM is that it is relatively straightforward, both conceptually and computationally, to examine a given spin-lattice model for different values of s, within a unified and consistent hierarchy of approximations. Hence, we now use the method to examine the J_1–J_2–J_3 model on the honeycomb lattice, along the line $J_3 = J_2 \equiv \kappa J_1 > 0$, with $0 \leq \kappa \leq 1$, for the cases $s = 1, \frac{3}{2}, 2, \frac{5}{2}$, in order to compare them both with the extreme quantum limit ($s = \frac{1}{2}$) case and with the classical ($s \to \infty$) case.

3. The coupled cluster method

We briefly describe here the key features of the CCM, and refer the interested reader to the extensive literature (and see, e.g., Refs. [43–50] and references cited therein) for further details. To implement the method in practice one first needs to choose a suitable normalized model (or reference) state $|\Phi\rangle$, against which the correlations present in the exact GS wave function can be incorporated. The properties required of $|\Phi\rangle$ are described more fully below, but in general terms it plays the role of a generalized vacuum state. For the present study suitable choices for the model state $|\Phi\rangle$ will turn out to be the two quasiclassical AFM states (viz., the Néel and collinear striped states) that form the stable GS phases of the classical version of the model under consideration in their respective regimes of the $T = 0$ phase diagram.

The exact GS ket- and bra-state wave functions, $|\Psi\rangle$ and $\langle\tilde{\Psi}|$, respectively, are chosen to have the normalization conditions,

$$\langle\tilde{\Psi}|\Psi\rangle = \langle\Phi|\Psi\rangle = \langle\Phi|\Phi\rangle \equiv 1\,. \tag{6}$$

These exact states are now parametrized with respect to the model state $|\Phi\rangle$ in the exponentiated forms,

$$|\Psi\rangle = e^{S}|\Phi\rangle\,; \quad \langle\tilde{\Psi}| = \langle\Phi|\tilde{S}e^{-S}\,, \tag{7}$$

that are a characteristic hallmark of the CCM. Although the correlation operator $\tilde{S}$ may formally be expressed in terms of its counterpart S as

$$\langle\Phi|\tilde{S} = \frac{\langle\Phi|e^{S^\dagger}e^{S}}{\langle\Phi|e^{S^\dagger}e^{S}|\Phi\rangle}\,, \tag{8}$$

by using Hermiticity, the CCM chooses not to restrain this relationship between $|\Psi\rangle$ and $\langle\tilde{\Psi}|$ explicitly. Instead, the two correlation operators S and $\tilde{S}$ are formally decomposed independently as

$$S = \sum_{I\neq 0}\mathcal{S}_I C_I^+\,; \quad \tilde{S} = 1 + \sum_{I\neq 0}\tilde{\mathcal{S}}_I C_I^-\,, \tag{9}$$

where $C_0^+ \equiv 1$ is defined to be the identity operator in the many-body Hilbert space, and where the set index I denotes a complete set of single-particle configurations for all N particles. What is required of $|\Phi\rangle$ and the set of (multiconfigurational) creation operators $\{C_I^+\}$ is that $|\Phi\rangle$ is a fiducial (or cyclic) vector with respect to these operators, i.e., as a generalized vacuum state. Explicitly we require that the set of states $\{C_I^+|\Phi\rangle\}$ form a complete basis for the ket-state Hilbert space, and that

$$\langle\Phi|C_I^+ = 0 = C_I^-|\Phi\rangle\,, \quad \forall I \neq 0\,, \tag{10}$$

where the destruction operators $C_I^- \equiv (C_I^+)^\dagger$. Lastly, and importantly, we require that all members of the complete set of operators $\{C_I^+\}$ are mutually commuting.

The rather general CCM paramerizations of Eqs. (7)–(10) have several immediate important consequences. While Hermiticity is not made explicit, and while the exact correlation operators S and $\tilde{S}$ will certainly fulfill Eq. (8), when approximations are made (e.g., by truncating the sums over configuration I in Eq. (9), as is usually done in practice), Hermiticity may be only approximately maintained. Against this loss, however, come several advantages, which usually far outweigh it. Firstly, the CCM parametrizations guarantee that the Goldstone linked cluster theorem is exactly preserved, as we describe in more detail below, even if truncations are made in Eq. (9). In turn, this feature guarantees size-extensivity at any such level of truncation, so that the GS energy, for example, is always calculated as an extensive variable. Thus, the CCM has the first advantage that we may work from the outset in the thermodynamic limit ($N \to \infty$), thereby obviating the need for any finite-size scaling, as is required in most alternative methods. A second key feature of the CCM, which is guaranteed by its exponentiated parametrizations, is that it also exactly preserves the important Hellmann-Feynman theorem at any level of truncation or approximate implementation.

Clearly, a knowledge of the CCM c-number correlation coefficients $\{\mathcal{S}_I, \tilde{\mathcal{S}}_I\}$ completely suffices to determine the GS expectation value of any operator. They are now found by minimization of the GS energy expectation functional,

$$\bar{H} = \bar{H}(\mathcal{S}_I, \tilde{\mathcal{S}}_I) \equiv \langle\Phi|\tilde{S}e^{-S}He^{S}|\Phi\rangle\,, \tag{11}$$

from Eq. (7), with respect to each of the coefficients $\{\mathcal{S}_I, \tilde{\mathcal{S}}_I\,; \forall I \neq 0\}$ separately. Variation of $\bar{H}$ from Eq. (11), with respect to $\tilde{\mathcal{S}}_I$ from Eq. (9), immediately yields

$$\langle\Phi|C_I^- e^{-S}He^{S}|\Phi\rangle = 0\,, \quad \forall I \neq 0\,, \tag{12}$$

which is a coupled set of *non-linear* equations for the coefficients $\{\mathcal{S}_I\}$, with the same number of equations as parameters. A similar variation of $\bar{H}$ from Eq. (11), with respect to $\mathcal{S}_I$ from Eq. (9) yields

$$\langle\Phi|\tilde{S}e^{-S}[H, C_I^+]e^S|\Phi\rangle = 0\,, \quad \forall I \neq 0\,, \tag{13}$$

as a coupled set of *linear* equations for the coefficients $\{\tilde{\mathcal{S}}_I\}$, again with the same number of equations as parameters, once the coefficients $\{\mathcal{S}_I\}$ are used as input after Eq. (12) has been solved for them.

The GS energy E, which is simply the value of $\bar{H}$ from Eq. (11) at the minimum, may then be expressed as

$$E = \langle\Phi|e^{-S}He^S|\Phi\rangle = \langle\Phi|He^S|\Phi\rangle\,, \tag{14}$$

using Eqs. (12) and (13). By making use of Eq. (14), we may rewrite the set of linear equations (13) in the equivalent form,

$$\langle\Phi|\tilde{S}(e^{-S}He^S - E)C_I^+|\Phi\rangle = 0\,, \quad \forall I \neq 0\,, \tag{15}$$

which is just a set of generalized linear eigenvalue equations for the set of coefficients $\{\tilde{\mathcal{S}}_I\}$.

Up to this point in the CCM procedure and implementation we have made no approximations. However, clearly Eqs. (12) that determine the creation coefficients $\{\mathcal{S}_I\}$ are intrinsically highly nonlinear in view of the exponential terms. Hence one may ask if we now need to make truncations to evaluate these terms. We note, though, that these always appear in the equations to be solved in the combination $e^{-S}He^S$ of a similarity transformation of the Hamiltonian. This may itself be expanded as the well-known nested commutator sum,

$$e^{-S}He^S = \sum_{n=0}^{\infty}\frac{1}{n!}[H, S]_n\,, \tag{16}$$

where $[H, S]_n$ is an n-fold nested commutator, defined iteratively as

$$[H, S]_n = [[H, S]_{n-1}, S]\,; \quad [H, S]_0 = H\,. \tag{17}$$

Another key feature of the CCM is that this otherwise infinite sum now (usually) terminates exactly at some finite order, when used in the equations to be solved, due to the facts that all of the terms in the expansion of Eq. (9) for S commute with one another and that H itself is (usually, as here) of finite order in the relevant single-particle operators. For example, if H contains up to m-body interactions, in its second-quantized form it contains sums of terms involving products of up to $2m$ single-particle (destruction and creation) operators, and the sum in Eq. (16) will terminate exactly with the term $n = 2m$. In our present case where the Hamiltonian of Eq. (1) is bilinear in the SU(2) spin operators, the sum terminates at $n = 2$. Finally, we also note here that the fact that all of the operators in the set $\{C_I^+\}$ that comprise S by Eq. (9) commute with one another, automatically guarantees that all (nonzero) terms in the sum of Eq. (16) are linked to the Hamiltonian. Unlinked terms simply cannot appear, and hence the Goldstone theorem and size-extensivity are satisfied, at any level of truncation.

Hence, for any implementation of the CCM, the *only* approximation made in practice is to restrict the set of multiconfigurational set-indices $\{I\}$ that are retained in the expansions of Eq. (9) for the correlation operators $\{S, \tilde{S}\}$ to some appropriate (finite or infinite) subset. How this choice is made must clearly depend on the problem at hand and on the particular choices that have been made for the model state $|\Phi\rangle$ and the associated set of operators $\{C_I^+\}$. Let us, therefore, now turn to how such choices are made for the present model in particular and for quantum spin-lattice models in general.

The simplest choice of model state $|\Phi\rangle$ for a quantum spin-lattice problem is a straightforward independent-spin product state in which the spin projection (along some specified quantization axis) of the spin on each lattice site is specified independently. The two quasiclassical collinear AFM states shown in Figs. 1(a) and 1(b), viz., the Néel and striped states, are examples. In order to treat all such states in the same way it is very convenient to make a passive rotation of each spin independently (i.e., by making a suitable choice of local spin quantization axes on each site independently), so that on every site the spin points downwards, say, in the negative z_s direction, as in the spin-coordinate frame shown in Fig. 1. Such rotations are clearly just unitary transformations that leave the basic SU(2) spin commutation relations unchanged. In this way each lattice site k is completely equivalent to all others, and all such independent-spin product model states now take the universal form $|\Phi\rangle = |\downarrow\downarrow\downarrow\cdots\downarrow\rangle$.

In this representation it is now clear that $|\Phi\rangle$ can indeed be regarded as a fiducial vector with respect to a set of mutually commuting creation operators $\{C_I^+\}$, which may now be chosen as a product of single-spin raising operators, $s_k^+ \equiv s_k^x + is_k^y$, such that $C_I^+ \to s_{k_1}^+ s_{k_2}^+ \cdots s_{k_n}^+$; $n = 1, 2, \cdots, 2sN$. The corresponding set index I thus becomes a set of lattice-site indices, $I \to \{k_1, k_2, \cdots, k_n;\ n = 1, 2, \cdots, 2sN\}$, in which each site index may be repeated up to $2s$ times. Once the local spin coordinates have been selected by the above procedure (i.e., for the given model state $|\Phi\rangle$), one simply re-expresses the Hamiltonian H in terms of them.

We now turn to the choice of approximation scheme, which hence simply involves a choice of which configurations $\{I\}$ to retain in the decompositions of Eq. (9) for the CCM correlation operators $(S, \tilde{S})$. A powerful and rather general such scheme, the so-called SUBn–m scheme, retains the configurations involving a maximum of n spin-flips (where each spin-flip requires the action of a spin-raising operator s_k^+ acting once) spanning a range of no more than m contiguous sites on the lattice. A set of lattice sites is defined to be contiguous if every site in the set is the NN of at least one other in the set (in a specified geometry). Clearly, as both indices become indefinitely large, the approximation becomes exact. Different schemes can be defined according to how each index approaches infinity.

For example, if we first let $m \to \infty$, we arrive at the so-called SUB$n \equiv$ SUBn–∞ scheme, which is the approximation scheme most commonly employed, more generally, for systems defined in a spatial continuum, such as atoms and molecules in quantum chemistry [51] or finite atomic nuclei or nuclear matter in nuclear physics [52] (and see, e.g., Refs. [43, 44] for further details). By contrast to continuum theories, for which the notion of contiguity is not easily applicable, in lattice theories both indices n and m may be kept finite. A very commonly used scheme is the so-called LSUBm scheme [50, 53], defined to retain, at the mth level of approximation, all spin clusters described by multispin configurations in the index set $\{I\}$ defined over any possible lattice animal (or polyomino) of size m on the lattice. Again, such a lattice animal is defined in the usual graph-theoretic sense to be a configured set of contiguous (in the above sense) sites on the lattice. Clearly, the LSUBm scheme is equivalent to the SUBn–m scheme when $n = 2sm$ for particles of spin quantum number s, i.e., LSUB$m \equiv$ SUB$2sm$–m. The LSUBm scheme was precisely the truncation scheme used in our previous study of the present model for the case $s = \frac{1}{2}$ [9].

At a given mth level of LSUBm approximation the number N_f of fundamental spin configurations that are distinct (under the symmetries of the lattice and the specified model state), which are retained is lowest for $s = \frac{1}{2}$ and rises sharply as s is increased. Since N_f typically also increases rapidly (typically, faster than exponentially) with the truncation index m, an alternative scheme for use in cases $s > \frac{1}{2}$ is to be set $m = n$ and hence employ the resulting SUBn–n scheme. Clearly, the two schemes are equivalent only for the case $s = \frac{1}{2}$, for which LSUB$m \equiv$ SUBm–m. We note too that the numbers N_f of fundamental configurations at a given SUBn–n level are still higher for the cases $s > \frac{1}{2}$ considered here than for the case $s = \frac{1}{2}$. Thus, whereas for the the present model we were able to perform LSUBm calculations with $m \leq 12$

previously for the case $s = \frac{1}{2}$ [9], we are now restricted for the cases $s > \frac{1}{2}$ considered here to perform SUBn–n calculations with $n \leq 10$, with similar amounts of supercomputer resources available. Thus, for example, for the case $s = \frac{1}{2}$, at the LSUB12 level of approximation we have $N_f = 103,097(250,891)$ using the Néel (striped) state as the CCM model state. By comparison, at the SUB10–10 level of approximation we have $N_f = 219,521(552,678)$ for the case $s = 1$, and $N_f = 538,570(1,436,958)$ for the case $s = \frac{5}{2}$, in each case using the Néel (striped) state as the CCM model state. Just as before [9] we employ massively parallel computing [54] both to derive (with computer algebra) and to solve (and see, e.g., Ref. [53]) the respective coupled sets of CCM equations (12) and (15).

Once the coefficients $\{\mathcal{S}_I, \tilde{\mathcal{S}}_I\}$ retained in a given SUBn–n approximation have been calculated by solving Eqs. (12) and (15), we may calculate any GS quantity at the same level of approximation. Thus, for example, the GS energy E may be calculated from Eq. (14) in terms of the ket-state coefficients $\{\mathcal{S}_I\}$ alone. Any other GS quantity requires a knowledge also of the bra-state coefficients $\{\tilde{\mathcal{S}}_I\}$. For example, we also calculate here the magnetic order parameter M, which is defined to be the average on-site GS magnetization,

$$M = -\frac{1}{N}\sum_{k=1}^{N}\langle\Phi|\tilde{S}e^{-S}s_k^z e^S|\Phi\rangle\,, \tag{18}$$

in terms of the local rotated spin-coordinate frames that we have described above.

As a last step, and as essentially the only approximation made in the whole CCM implementation, we need to extrapolate the raw SUBn–n data points for E and M to the exact $n \to \infty$ limit. Although no exact extrapolation rules are known, a great deal of experience has by now been accumulated for doing so, from the many applications of the technique to a wide variety of spin-lattice problems that have been examined with the method. For the GS energy per spin, for example, a very well tested and highly accurate extrapolation ansatz (and see, e.g., Refs. [9, 15–19, 50, 55–67]) is

$$\frac{E(n)}{N} = a_0 + a_1 n^{-2} + a_2 n^{-4}\,, \tag{19}$$

while for the magnetic order parameter M different schemes have been used in different situations. Unsurprisingly, the GS expectation values of other physical observables generally converge less rapidly than the GS energy, i.e., with leading exponents less than two. More specifically, the leading exponent for M tends to depend on the amount of frustration present, generally being smaller for the most highly frustrated cases.

Thus, for unfrustrated models or for models with only moderate amounts of frustration present, a scaling ansatz for $M(n)$ with leading power $1/n$ (rather than $1/n^2$ as for the GS energy),

$$M(n) = b_0 + b_1 n^{-1} + b_2 n^{-2}\,, \tag{20}$$

has been found to work well in many cases (and see, e.g., Refs. [15–17, 19, 55–58, 62–66]). For systems that are either close to a QCP or for which the magnetic order parameter M for the phase under study is either zero or close to zero, the extrapolation ansatz of Eq. (20) tends to overestimate the extrapolated value and hence to predict a somewhat too large value for the critical strength of the frustrating interaction that is driving the respective phase transition. In such cases a great deal of evidence has now shown that a scaling ansatz with leading power $1/n^{1/2}$ fits the SUBn–n data much better. Thus, as an alternative in those instances to Eq. (20), a more appropriate scaling scheme (and see, e.g., Refs. [9, 15–19, 59–61, 67]) is

$$M(n) = c_0 + c_1 n^{-1/2} + c_2 n^{-3/2}\,. \tag{21}$$

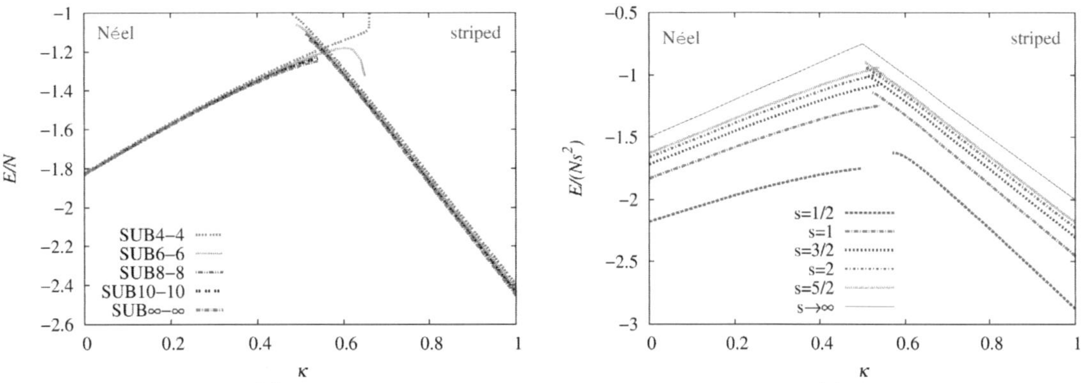

(a) (b)

Figure 2. CCM results for the GS energy per spin, E/N, for the J_1–J_2–J_3 model on the honeycomb lattice (with $J_1 \equiv 1, J_3 = J_2 \equiv \kappa J_1 > 0$), as a function of κ, using the Néel and striped states as the CCM model states. (a) The results for the $s = 1$ model are shown using the SUBn–n approximations with $n = \{4, 6, 8, 10\}$, together with the corresponding extrapolated SUB∞–∞ results obtained using equation (19), with this data set. (b) We show extrapolated (SUB∞–∞) results for $E/(Ns^2)$ as a function of κ, using equation (19) together with the data set $n = \{6, 8, 10, 12\}$ for the case $s = \frac{1}{2}$, and the corresponding data sets with $n = \{4, 6, 8, 10\}$ for the four cases $s = 1, \frac{3}{2}, 2, \frac{5}{2}$. We also show the corresponding classical ($s \to \infty$) result.

Since the extrapolation schemes of Eqs. (19)–(21) contain three fitting parameters, it is clearly preferable to use at least four SUBn–n data points in each case. Furthermore since the lowest-order SUB2–2 approximants are less likely to conform well to the extrapolation schemes, we prefer to perform fits using SUBn–n data with with $n \geq 4$.

4. Results

We show in Fig. 2(a) our CCM results for the GS energy per spin, E/N, of the spin-1 model at various SUBn–n levels of approximation with $n = \{4, 6, 8, 10\}$, using both the Néel and striped AFM states as separate choices of the CCM model state. We observe very clearly that the results converge very rapidly with increasing values of the truncation index n, and we also show the extrapolated $n \to \infty$ results, a_0, from Eq. (19). We also observe that each of the energy curves based on a particular model state terminates at a critical value of the frustration parameter κ that depends on the SUBn–n approximation used. Beyond those critical values no real solutions can be found to the corresponding CCM equations (12). Such termination points of the CCM coupled equations are very common in practice, and are well understood (see, e.g., Refs. [16, 50, 62]). They are direct manifestations of the corresponding QCP in the system, at which the respective form of magnetic LRO in the model state used melts. As is usually the case, the CCM SUBn–n solutions for a given finite value of n and for a given phase extend beyond the actual SUB∞–∞ QCP, i.e., into the unphysical regime beyond the termination point. The extent of the unphysical regime diminishes (to zero) as the truncation order n increases (to the exact $n \to \infty$ limit).

In Fig. 2(b) we compare the corresponding extrapolated curves for the scaled GS energy per spin, $E/(Ns^2)$, using both the Néel and striped AFM states separately as our choice of CCM model state, for the five cases $s = \frac{1}{2}, 1, \frac{3}{2}, 2, \frac{5}{2}$. In each case the extrapolation

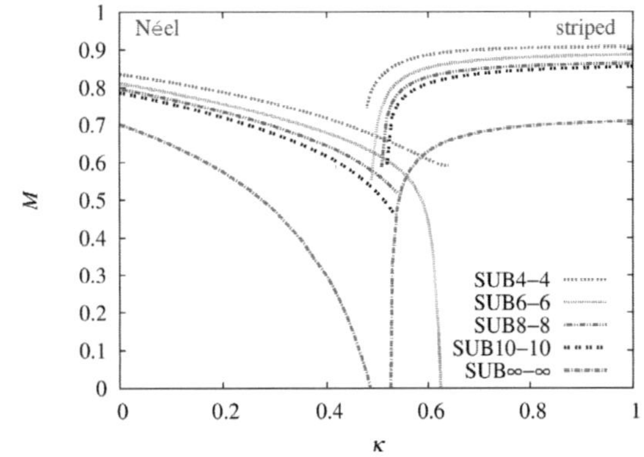

(a)

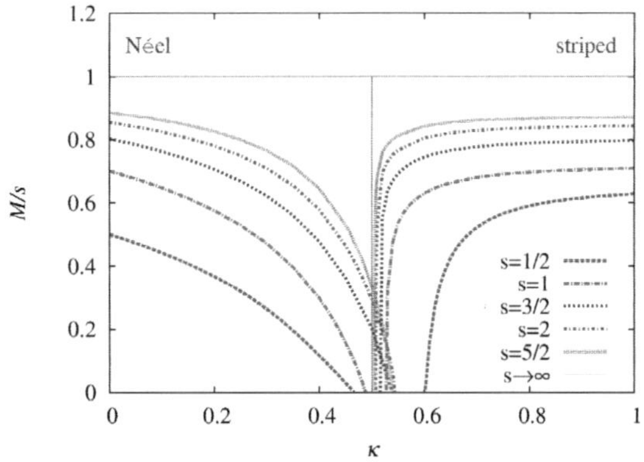

(b)

Figure 3. CCM results for the GS magnetic order parameter, M, for the J_1–J_2–J_3 model on the honeycomb lattice (with $J_1 \equiv 1, J_3 = J_2 \equiv \kappa J_1 > 0$), as a function of κ, using the Néel and striped states as the CCM model states. (a) The results for the $s = 1$ model are shown using the SUBn–n approximations with $n = \{4, 6, 8, 10\}$, together with the corresponding extrapolated SUB∞–∞ results obtained using equation (21), with this data set. (b) We show extrapolated (SUB∞–∞) results for M/s as a function of κ, using equation (21) together with the data set $n = \{6, 8, 10, 12\}$ for the case $s = \frac{1}{2}$, and the corresponding data sets with $n = \{4, 6, 8, 10\}$ for the four cases $s = 1, \frac{3}{2}, 2, \frac{5}{2}$. We also show the corresponding classical ($s \to \infty$) result.

is performed with Eq. (19). For the case $s = \frac{1}{2}$ alone the input SUBn–n data points are $n = \{6, 8, 10, 12\}$, while for each of the cases $s > \frac{1}{2}$ the input set is $n = \{4, 6, 8, 10\}$. We also show in Fig. 2(b) the corresponding classical ($s \to \infty$) results, $E_{\rm cl}^{\rm N\acute{e}el}/(Ns^2) = \frac{3}{2}(-1 + \kappa)$ and $E_{\rm cl}^{\rm striped}/(Ns^2) = \frac{1}{2}(1 - 5\kappa)$. We observe clear preliminary evidence from Fig. 2(b) for an intermediate phase (between the phases with Néel and striped magnetic LRO) in the $s = 1$ case, although with a range of stability in the frustration parameter κ now markedly less than in the $s = \frac{1}{2}$ case. The preliminary evidence from the energy results is also that there is no such intermediate phase present in each of the cases $s > 1$. Lastly, Fig. 2(b) also shows that, at least so far as the energy results are concerned, all cases with $s \gtrsim 2$ are rather close to the classical limit.

In order to get more detailed evidence on the phase structures of the model for various values of the spin quantum number s we now turn to the results for the GS magnetic order parameter, M, of Eq. (18). Thus, firstly, in Fig. 3(a) we show our CCM results for M for the spin-1 model at various SUBn–n levels of approximation with $n = \{4, 6, 8, 10\}$, using both the Néel and striped AFM states as separate choices for the model state. Hence, what is shown in Fig. 3(a) for M is just the precise analogue of what is shown in Fig. 2(a) for E/N. It is clear that the SUBn–n sequence of approximations for M converges more slowly than for E/N, just as expected. We also show in Fig. 3(a) the extrapolated results for the spin-1 model, where we have used the data set shown, $n = \{4, 6, 8, 10\}$, as input to the extrapolation scheme of Eq. (21). As was explained in Sec. 3, while the alternative scheme of Eq. (20) is certainly more appropriate when the frustration parameter κ is zero or small, that of Eq. (21) is certainly preferable for larger values of κ (e.g., in the striped phase) or when the order parameter M becomes small (i.e., near any QCPs).

The SUB∞–∞ extrapolation shown in Fig. 3(a) now clearly validates the earlier, more qualitative, results from the GS energy, namely the existence of a GS phase intermediate between the quasiclassical Néel and striped collinear AFM states, just as in the spin-$\frac{1}{2}$ case. Once again, Néel LRO exists over the range $0 \leq \kappa < \kappa_{c_1}$, while striped LRO exists for $\kappa > \kappa_{c_2}$, where $\kappa_{c_1} < \kappa_{\rm cl} = \frac{1}{2}$ and $\kappa_{c_2} > \kappa_{\rm cl} = \frac{1}{2}$. The values obtained for the two QCPs from the extrapolations using Eq. (21) with the data set $n = \{4, 6, 8, 10\}$, taken as the points where $M \rightarrow 0$, as shown in Fig. 3(a), are $\kappa_{c_1} \approx 0.486$ and $\kappa_{c_2} \approx 0.527$. A more detailed analysis of the errors associated with the fits, and by comparison with comparable extrapolations using alternative data sets (e.g., $n = \{6, 8, 10\}$ and $n = \{4, 6, 8\}$), yields our best estimates for the spin-1 model QCPs, $\kappa_{c_1} = 0.485(5)$ and $\kappa_{c_2} = 0.528(5)$. These may be compared with the corresponding values for the spin-$\frac{1}{2}$ model QCPs [9], $\kappa_{c_1} = 0.47$ and $\kappa_{c_2} = 0.60$.

In Fig. 3(b) we now compare the corresponding extrapolated curves for the scaled magnetic order parameter, M/s, using both the Néel and striped AFM states separately as CCM model states, for the five cases $s = \frac{1}{2}, 1, \frac{3}{2}, 2, \frac{5}{2}$. In each case shown the extrapolation has been performed with the ansatz of Eq. (21), together with the data set $n = \{6, 8, 10, 12\}$ for the $s = \frac{1}{2}$ case and the sets $n = \{4, 6, 8, 10\}$ for each case with $s > \frac{1}{2}$. The results once again validate our earlier, more qualitative, findings from the GS energy results, that the intermediate phase is present only for the two cases $s = \frac{1}{2}$, 1, with a direct transition from the Néel to the striped phase in all cases $s > 1$, just as in the classical ($s \rightarrow \infty$) limit. For all cases $s > 1$ this direct transition clearly occurs at values very close to the classical value $\kappa_{\rm cl} = \frac{1}{2}$. The actual crossing points of the order parameter curves shown in Fig. 3(b) occur at values $\kappa_m \approx 0.517$ for $s = \frac{3}{2}$, $\kappa_m \approx 0.508$ for $s = 2$, and $\kappa_m \approx 0.505$ for $s = \frac{5}{2}$. What is apparent from Fig. 3(b) is that the curves for the striped phase approach zero much more steeply than for the Néel phase for all values of s. For the case $s = \frac{1}{2}$ it was argued [9] that this was a reflection of the transition at κ_{c_2} being of first-order type, while that at κ_{c_1} is of continuous (and hence probably of the deconfined) type. The difference in the shapes of the curves near the crossing point is what leads to the direct transition apparently being at values slightly larger than $\kappa_{\rm cl} = \frac{1}{2}$ for finite values of $s > 1$.

Clearly the precise crossing points κ_m of the magnetic order curves for the Néel and striped phases for the cases $s > 1$ depend rather critically on the extrapolations, particularly those for the striped phase, where the slope become large. In such cases more precise values of the corresponding QCP for the direct transition between the two quasiclassical phases can be expected to come from the analogous crossing points, κ_e, of the extrapolated energy curves. The respective values from Fig. 2(b) are $\kappa_e \approx 0.544$ for $s = \frac{3}{2}$, $\kappa_e \approx 0.534$ for $s = 2$, and $\kappa_e \approx 0.528$ for $s = \frac{5}{2}$. It is reassuring that the respective pairs of values of κ_e and κ_m agree so well in each case, for what are essentially quite independent results.

Before discussing how we can investigate the nature of the intermediate phase for the present $s = 1$ case within the CCM framework, let us briefly comment on the case of the pure HAF on the honeycomb lattice, with NN interactions only (i.e., the limiting case $\kappa = 0$ of the present model). In this case, the extrapolation ansatz of Eq. (20) becomes applicable, rather than that of Eq. (21) shown in Fig. 3. We show in Table 1 the scaled values for the GS energy per spin and magnetic order parameters, $E/(Ns^2)$ and M/s, respectively, for our present model calculations at the unfrustrated limiting value $\kappa = 0$. The corresponding extrapolation schemes of Eqs. (19) and Eq. (20) have been used in Table 1, together with the input data sets with $n = \{6, 8, 10, 12\}$ for $s = \frac{1}{2}$, and with $n = \{4, 6, 8, 10\}$ for $s > \frac{1}{2}$.

Another way to estimate the accuracy of our extrapolated CCM results for the higher spin values is to use them to extract, for example, the coefficients of the expansions of $E/(Ns^2)$ and M/s in inverse powers of s, and compare them with the results of higher-order spin-wave theory (SWT). For example, at second-order, we may fit our results of Table 1 to the forms,

$$\frac{E}{Ns^2} = -\frac{3}{2} + \frac{e_1}{s} + \frac{e_2}{s^2}, \qquad (22)$$

Journal of Physics: Conference Series **702** (2016) 012001 doi:10.1088/1742-6596/702/1/012001

Table 1. GS parameters of the HAF on the honeycomb lattice, with NN interactions only (of strength $J_1 = 1$), for various values of the spin quantum number s.

s	$E/(Ns^2)$	M/s
$\frac{1}{2}$	-2.17866	0.5459
1	-1.83061	0.7412
$\frac{3}{2}$	-1.71721	0.8249
2	-1.66159	0.8689
$\frac{5}{2}$	-1.62862	0.8955
∞	-1.5	1

and

$$\frac{M}{s} = 1 + \frac{m_1}{s} + \frac{m_2}{s^2}, \tag{23}$$

and then compare with the corresponding results of second-order SWT, i.e., SWT(2). If we simply take our results from Table 1 for the two highest spin values calculated, viz., $s = 2, \frac{5}{2}$, and fit them to Eqs. (22) and (23), we obtain values $e_1 \approx -0.31503$ and $e_2 \approx -0.01630$ for the GS energy, and $m_1 \approx -0.2575$ and $m_2 \approx -0.0095$ for the GS Néel magnetic order parameter (i.e., the sublattice magnetization). The corresponding (exact) SWT(2) results [68] are $e_1 = -0.31476$, $e_2 = -0.01651$, $m_1 = -0.2582$, and $m_2 = 0$. The agreement is rather striking.

We now turn finally to the question of what is the nature of the intermediate phase in the case $s = 1$. For the analogous $s = \frac{1}{2}$ case it was shown [9] that the intermediate paramagnetic phase likely had PVBC order. It is natural now to consider this possibility for the $s = 1$ case. To do so we now calculate within the CCM framework the susceptibility, χ_p, which measures the response of the system to an applied external field that promotes PVBC order. More generally, let us add an infinitesimal field operator $F \equiv \delta\,\hat{O}$ to the Hamiltonian H of Eq. (1). We then calculate the perturbed energy per site, $E(\delta)/N = e(\delta)$, for the perturbed Hamiltonian $H + F$, using the same CCM procedure as above, and using the same previous model states. The susceptibility of the system to the perturbed operator $\hat{O}$ is then defined as usual to be

$$\chi_F \equiv -\left.\frac{\partial^2 e(\delta)}{\partial \delta^2}\right|_{\delta=0}, \tag{24}$$

so that the energy,

$$e(\delta) = e(0) - \frac{1}{2}\chi_F \delta^2, \tag{25}$$

is a maximum at $\delta = 0$ for $\chi_F > 0$. A clear signal of the system becoming unstable against the perturbation F is the finding that χ_F diverges or, equivalently, that χ_F^{-1} becomes zero (and then possibly changes sign).

In our present case the perturbing operator F is now chosen to promote PVBC order, and it is illustrated in Fig. 4(a). It clearly breaks the translational symmetry of the system. The SUBn–n estimates, $\chi_p(n)$, for the resulting susceptibility of our system to the formation of PVBC order, then need to be extrapolated to the exact ($n \to \infty$) limit. Previous experience [9] has shown that an appropriate extrapolation ansatz is

$$\chi_p^{-1}(n) = x_0 + x_1 n^{-2} + x_2 n^{-4}. \tag{26}$$

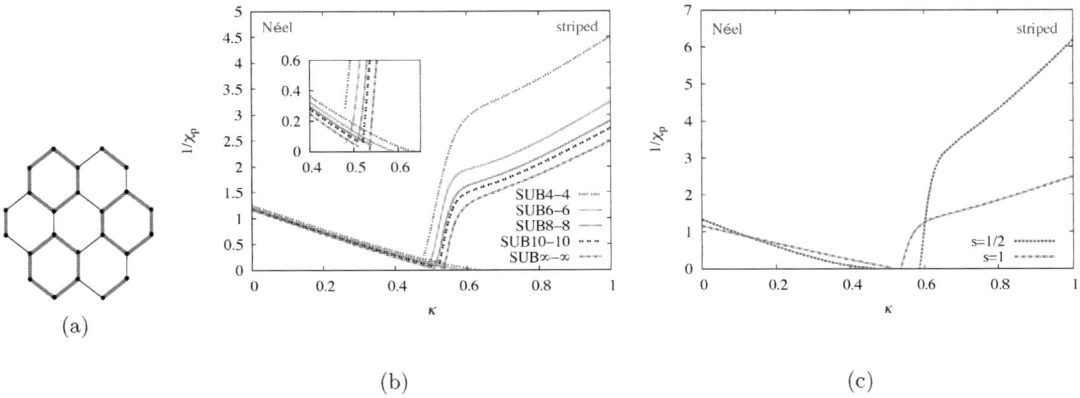

Figure 4. (a) The fields $F = \delta\ \hat{O}$ for the plaquette susceptibility χ_p. Thick (red) and thin (black) lines correspond respectively to strengthened and weakened NN exchange couplings, where $\hat{O} = \sum_{\langle i,j\rangle} a_{ij}\mathbf{s}_i \cdot \mathbf{s}_j$, and the sum runs over all NN bonds, with $a_{ij} = +1$ and -1 for thick (red) and thin (black) lines respectively. (b) CCM results for the inverse plaquette susceptibility, $1/\chi_p$, for the J_1–J_2–J_3 model on the honeycomb lattice (with $J_1 \equiv 1, J_3 = J_2 \equiv \kappa J_1 > 0$), as a function of κ, using the Néel and striped states as the CCM model states. The results for the $s = 1$ model are shown using the SUBn–n approximations with $n = \{4, 6, 8, 10\}$, together with the corresponding extrapolated SUB∞–∞ results obtained using equation (26), with this data set. (c) We show extrapolated (SUB∞–∞) results for $1/\chi_p$ as a function of κ, using equation (26) together with the data set $n = \{6, 8, 10, 12\}$ for the case $s = \frac{1}{2}$, and the data set $n = \{4, 6, 8, 10\}$ for the case $s = 1$.

We show in Fig. 4(b) our CCM results for $\chi_p^{-1}(n)$ at SUBn–n levels of approximation with $n = \{4, 6, 8, 10\}$, using both the Néel and striped AFM states separately as model states, in complete analogy to what is displayed in Figs. 2(a) and 3(a) for the GS energy per spin, E/N, and magnetic order parameter, M. In Fig. 4(b) we also show the extrapolated results (x_0) obtained by inserting the set of raw results shown into Eq. (26). Just as in the spin-$\frac{1}{2}$ case [9], the results for $\chi_p^{-1}(n)$ converge much faster for the Néel state than for the striped state. For both states there are clear critical points at which χ_p^{-1} vanishes. However, the shapes of the curves for χ_p^{-1} near their respective critical points differ markedly, just as in the spin-$\frac{1}{2}$ case. Thus, on the Néel side, $\chi_p^{-1} \to 0$ with a slope that is small. By contrast, on the striped side, $\chi_p^{-1} \to 0$ with a very large slope (and which is probably compatible with being infinite, within extrapolation errors). These differences reinforce our earlier findings that the critical point at which Néel order vanishes is likely to mark a continuous phase transition, while that at which striped order vanishes is likely to mark a first-order transition.

In Fig. 4(c) we compare the extrapolated CCM results for χ_p^{-1} for the two cases ($s = \frac{1}{2}$, 1) for which our findings indicate the existence in the model of a phase in its $T = 0$ GS phase diagram intermediate between the two quasiclassical phases with magnetic LRO. Whereas in the spin-$\frac{1}{2}$ case there exists a clear gap along the frustration parameter, κ, axis between the two points at which χ_p^{-1} vanishes (one for each quasiclassical phase), the gap in the spin-1 case is much less marked. Indeed, its very existence is open to doubt, as we explain below. For the spin-$\frac{1}{2}$ case [9] the two values on Fig. 4(c) at which $\chi_p^{-1} \to 0$ are $\kappa \approx 0.473$ and $\kappa \approx 0.586$.

These may be compared with the corresponding values on Fig. 3(b) at which $M \to 0$, which are $\kappa_{c_1} \approx 0.466$ and $\kappa_{c_2} \approx 0.601$. The very close agreement between the corresponding values was taken [9] to be good evidence that the PVBC phase occurs at (or is very close to) the transition points κ_{c_1} and κ_{c_2} where the quasiclassical magnetic LRO vanishes. The fact that the slope of the $\chi_p^{-1}(\kappa)$ curve on the Néel side is vanishingly small (within numerical errors) at the point where $\chi_p^{-1} \to 0$, also provided strong evidence that χ_p^{-1} vanishes over the entire intermediate region in Fig. 4(c) between the points, where the CCM calculations have been performed with the two classical model states. All of this evidence pointed strongly to the stable GS phase in the whole of the region $\kappa_{c_1} < \kappa < \kappa_{c_2}$ being one with PVBC order, in the $s = \frac{1}{2}$ case. If any other phase exists in part of this region, its region of stability is clearly constrained by the CCM calculations [9] to be a very small part of the intermediate region.

The corresponding situation for the $s = 1$ case is now subtly different, however. Thus, firstly, a close inspection of Fig. 4(b) (and, especially, the inset) shows that on the Néel side the extrapolated SUB∞–∞ $\chi_p^{-1}(\kappa)$ curve has not quite reached zero at the SUB10–10 termination point (or at least as far as we have managed to perform numerical calculations, which, as we discussed above, become increasingly difficult and computationally costly the closer one approaches a termination point). A simple further extrapolation of the curve, however, yields a value $\kappa \approx 0.530(2)$ at which $\chi_p^{-1} \to 0$ on the Néel side. This is very close to the corresponding value in Fig. 4(b) of $\kappa \approx 0.535$ at which $\chi_p^{-1} \to 0$ on the striped phase side. Significantly, both of these values are indubitably greater than $\kappa_{\rm cl} = \frac{1}{2}$. By contrast, the corresponding values from Fig. 3(b) at which $M \to 0$ for the spin-1 model are $\kappa_{c_1} \approx 0.486$ and $\kappa_{c_2} \approx 0.527$. The most likely interpretation of our χ_p results is hence that the CCM using both model states is showing that χ_p^{-1} vanishes only at a single point, viz., the QCP κ_{c_2}, for the spin-1 model. This interpretation is lent further weight by the observation that, unlike in the spin-$\frac{1}{2}$ case, the slope of the $\chi_p^{-1}(\kappa)$ curve on the Néel side does not appear to be zero at the (extrapolated) point at which it becomes zero.

In summary, if a PVBC-ordered phase is stable anywhere in the intermediate regime $\kappa_{c_1} < \kappa < \kappa_{c_2}$, our findings are that it is confined only to a very narrow range close to κ_{c_2}, and that it is definitely not the stable GS phase over the whole interval. The more likely scenario is that the intermediate regime is occupied by a paramagnetic phase (or more than one such phase) with a form of order other than PVBC.

5. Conclusions

We have investigated higher-spin versions of a frustrated J_1–J_2–J_3 HAF model on the honeycomb lattice, in the specific case where $J_1 > 0$, $J_3 = J_2 \equiv \kappa J_1 > 0$, over the range $0 \leq \kappa \leq 1$ of the frustration parameter. This includes the point of maximum frustration in the classical ($s \to \infty$) limit, viz., the tricritical point at $\kappa_{\rm cl} = \frac{1}{2}$, at which there is a direct first-order transition (along the line $J_3 = J_2$) between a Néel-ordered AFM phase for $\kappa < \kappa_{\rm cl}$ and a collinear stripe-ordered AFM phase for $\kappa > \kappa_{\rm cl}$. Whereas the spin-$\frac{1}{2}$ version of the model has been studied previously, higher-spin versions have received no attention, so far as we are aware.

In particular, the CCM has been applied to the spin-$\frac{1}{2}$ model in an earlier study [9] that yielded accurate results for its entire $T = 0$ GS phase diagram. Since the method generally provides values for the QCPs of a wide range of spin-lattice systems, which are among the most accurate available by any alternative methodology, we have now used the CCM to study spin-s versions of the model with values of the spin quantum number $s > \frac{1}{2}$. An aim has been to compare the $T = 0$ GS phase diagrams of the higher-spin models with the two extreme limits, $s = \frac{1}{2}$ (where quantum effects should be greatest) and $s \to \infty$ (where quantum effects vanish). In particular, it has been shown [9] that the direct classical transition at $\kappa_{\rm cl} = \frac{1}{2}$ is split by quantum fluctuations in the spin-$\frac{1}{2}$ model into two separate transitions at $\kappa_{c_1} < \kappa_{\rm cl}$

and $\kappa_{c_2} > \kappa_{c1}$, at the first of which Néel AFM order of the type shown in Fig. 1(a) breaks down, and at the second of which striped AFM order of the type shown in Fig. 1(b) breaks down. Between the regimes $\kappa < \kappa_{c_1}$ in which the stable GS phase has Néel magnetic LRO and $\kappa > \kappa_{c_2}$ in which the stable GS phase has striped magnetic LRO, there opens an intermediate paramagnetic regime $\kappa_{c_1} < \kappa < \kappa_{c_2}$. Strong evidence was presented [9] that the stable GS phase in this entire intermediate regime in the spin-$\frac{1}{2}$ case is an intrinsically quantum-mechanical one with PVBC order.

We have now performed analogous CCM calculations for spin-s versions of the same honeycomb lattice model for values $s = 1, \frac{3}{2}, 2, \frac{5}{2}$. A primary finding is that an intermediate phase also exists for the case $s = 1$, but that for all higher spins ($s \geq \frac{3}{2}$) the intermediate phase disappears in favour of a direct transition between the two quasiclassical states with magnetic LRO. For all finite values of the spin quantum number $s \geq \frac{3}{2}$ the direct transition seems to occur at a value κ_c marginally higher than the classical value of 0.5 [e.g., for $s = \frac{3}{2}$, $\kappa_c \approx 0.53(1)$, and for $s = \frac{5}{2}$, $\kappa_c \approx 0.52(1)$], with κ_c tending monotonically to 0.5 as $s \to \infty$.

The range of the intermediate phase is smaller for the spin-1 model ($\kappa_{c_1} \approx 0.49, \kappa_{c_2} \approx 0.53$) than for its spin-$\frac{1}{2}$ counterpart ($\kappa_{c_1} \approx 0.47, \kappa_{c_2} \approx 0.60$), as expected. Interestingly, all of the evidence garnered here is that, unlike in the spin-$\frac{1}{2}$ case the intermediate phase for the spin-1 model does *not* have PVBC ordering. On the other hand, both the spin-$\frac{1}{2}$ and spin-1 models seem to share that the transition at κ_{c_2} is a direct first-order one while that at κ_{c_1} is continuous.

On the basis of the present calculations, the nature and properties of the intermediate phase in the spin-1 version of the model remain open questions. It will be of considerable interest to study this phase further, both by the CCM and by the use of alternative techniques. A particularly promising such alternative technique in this respect is the DMRG method, which has been used very recently [25] in an analysis of the quantum ($T = 0$) phase diagram of the spin-1 version of the related J_1–J_2 Heisenberg model on the honeycomb lattice.

Acknowledgments

We thank the University of Minnesota Supercomputing Institute for the grant of supercomputing facilities. One of us (RFB) also gratefully acknowledges the Leverhulme Trust for the award of an Emeritus Fellowship (EM-2015-07).

References

[1] Mermin N D and Wagner H 1966 *Phys. Rev. Lett.* **17** 1133
[2] Rastelli E, Tassi A and Reatto L 1979 *Physica B & C* **97** 1
[3] Fouet J B, Sindzingre P and Lhuillier C 2001 *Eur. Phys. J. B* **20** 241
[4] Cabra D C, Lamas C A and Rosales H D 2011 *Phys. Rev. B* **83** 094506
[5] Mattsson A, Fröjdh P and Einarsson T 1994 *Phys. Rev. B* **49** 3997–4002
[6] Mulder A, Ganesh R, Capriotti L and Paramekanti A 2010 *Phys. Rev. B* **81** 214419
[7] Ganesh R, Sheng D N, Kim Y J and Paramekanti A 2011 *Phys. Rev. B* **83** 144414
[8] Clark B K, Abanin D A and Sondhi S L 2011 *Phys. Rev. Lett.* **107** 087204
[9] Farnell D J J, Bishop R F, Li P H Y, Richter J and Campbell C E 2011 *Phys. Rev. B* **84** 012403
[10] Reuther J, Abanin D A and Thomale R 2011 *Phys. Rev. B* **84** 014417
[11] Albuquerque A F, Schwandt D, Hetényi B, Capponi S, Mambrini M and Läuchli A M 2011 *Phys. Rev. B* **84** 024406
[12] Mosadeq H, Shahbazi F and Jafari S A 2011 *J. Phys.: Condens. Matter* **23** 226006
[13] Oitmaa J and Singh R R P 2011 *Phys. Rev. B* **84** 094424
[14] Mezzacapo F and Boninsegni M 2012 *Phys. Rev. B* **85** 060402(R)
[15] Li P H Y, Bishop R F, Farnell D J J, Richter J and Campbell C E 2012 *Phys. Rev. B* **85** 085115
[16] Bishop R F, Li P H Y, Farnell D J J and Campbell C E 2012 *J. Phys.: Condens. Matter* **24** 236002
[17] Bishop R F and Li P H Y 2012 *Phys. Rev. B* **85** 155135
[18] Li P H Y, Bishop R F, Farnell D J J and Campbell C E 2012 *Phys. Rev. B* **86** 144404
[19] Bishop R F, Li P H Y and Campbell C E 2013 *J. Phys.: Condens. Matter* **25** 306002
[20] Ganesh R, van den Brink J and Nishimoto S 2013 *Phys. Rev. Lett.* **110** 127203

[21] Zhu Z, Huse D A and White S R 2013 *Phys. Rev. Lett.* **111** 257201
[22] Zhang H and Lamas C A 2013 *Phys. Rev. B* **87** 024415
[23] Gong S S, Sheng D N, Motrunich O I and Fisher M P A 2013 *Phys. Rev. B* **88** 165138
[24] Yu X L, Liu D Y, Li P and Zou L J 2014 *Physica E* **59** 41
[25] Gong S S, Zhu W and Sheng D N 2015 Quantum phase diagram of the spin-1 J_1–J_2 Heisenberg model on the honeycomb lattice arXiv:1508.00515
[26] Miura Y, Hirai R, Kobayashi Y and Sato M 2006 *J. Phys. Soc. Jpn.* **75** 084707
[27] Kataev V, Möller A, Löw U, Jung W, Schittner N, Kriener M and Freimuth A 2005 *J. Magn. Magn. Mater.* **290–291** 310–313
[28] Tsirlin A A, Janson O and Rosner H 2010 *Phys. Rev. B* **82** 144416
[29] Climent-Pascual E, Norby P, Andersen N, Stephens P, Zandbergen H, Larsen J and Cava R 2012 *Inorg. Chem.* **51** 557–565
[30] Singh Y and Gegenwart P 2010 *Phys. Rev. B* **82** 064412
[31] Liu X, Berlijn T, Yin W G, Ku W, Tsvelik A, Kim Y J, Gretarsson H, Singh Y, Gegenwart P and Hill J P 2011 *Phys. Rev. B* **83** 220403(R)
[32] Singh Y, Manni S, Reuther J, Berlijn T, Thomale R, Ku W, Trebst S and Gegenwart P 2012 *Phys. Rev. Lett.* **108** 127203
[33] Choi S K, Coldea R, Kolmogorov A N, Lancaster T, Mazin I I, Blundell S J, Radaelli P G, Singh Y, Gegenwart P, Choi K R, Cheong S W, Baker P J, Stock C and Taylor J 2012 *Phys. Rev. Lett.* **108** 127204
[34] Regnault L P and Rossat-Mignod J 1990 *Magnetic properties of layered transition metal compounds* ed De Jongh L J (Dordrecht: Kluwer Academic Publishers) pp 271–321
[35] Roudebush J H, Andersen N H, Ramlau R, Garlea V O, Toft-Petersen R, Norby P, Schneider R, Hay J N and Cava R J 2013 *Inorg. Chem.* **52** 6083–6095
[36] Smirnova O, Azuma M, Kumada N, Kusano Y, Matsuda M, Shimakawa Y, Takei T, Yonesaki Y and Kinomura N 2009 *J. Am. Chem. Soc.* **131** 8313–8317
[37] Okubo S, Elmasry F, Zhang W, Fujisawa M, Sakurai T, Ohta H, Azuma M, Sumirnova O A and Kumada N 2010 *J. Phys.: Conf. Ser.* **200** 022042
[38] Villain J 1977 *J. Phys. (France)* **38** 385
[39] Villain J, Bidaux R, Carton J P and Conte R 1980 *J. Phys. (France)* **41** 1263
[40] Shender E F 1982 *Zh. Eksp. Teor. Fiz.* **83** 326–337
[41] Senthil T, Vishwanath A, Balents L, Sachdev S and Fisher M P A 2004 *Science* **303** 1490
[42] Senthil T, Balents L, Sachdev S, Vishwanath A and Fisher M P A 2004 *Phys. Rev. B* **70** 144407
[43] Bishop R F and Lührmann K H 1978 *Phys. Rev. B* **17** 3757–3780
[44] Bishop R F and Lührmann K H 1982 *Phys. Rev. B* **26** 5523–5557
[45] Arponen J 1983 *Ann. Phys. (N.Y.)* **151** 311–382
[46] Bishop R F and Kümmel H G 1987 *Phys. Today* **40(3)** 52
[47] Arponen J S and Bishop R F 1991 *Ann. Phys. (N.Y.)* **207** 171
[48] Bishop R F 1991 *Theor. Chim. Acta* **80** 95
[49] Bishop R F 1998 *Microscopic Quantum Many-Body Theories and Their Applications* Lecture Notes in Physics Vol. 510 ed Navarro J and Polls A (Berlin: Springer-Verlag) p 1
[50] Farnell D J J and Bishop R F 2004 *Quantum Magnetism* Lecture Notes in Physics Vol. 645 ed Schollwöck U, Richter J, Farnell D J J and Bishop R F (Berlin: Springer-Verlag) p 307
[51] Bartlett R J 1989 *J. Phys. Chem.* **93** 1697
[52] Kümmel H, Lührmann K H and Zabolitzky J G 1978 *Phys Rep.* **36C** 1
[53] Zeng C, Farnell D J J and Bishop R F 1998 *J. Stat. Phys.* **90** 327
[54] We use the program package CCCM of D. J. J. Farnell and J. Schulenburg, see http://www-e.uni-magdeburg.de/jschulen/ccm/index.html
[55] Bishop R F, Farnell D J J, Krüger S E, Parkinson J B, Richter J and Zeng C 2000 *J. Phys.: Condens. Matter* **12** 6887
[56] Krüger S E, Richter J, Schulenburg J, Farnell D J J and Bishop R F 2000 *Phys. Rev. B* **61** 14607
[57] Farnell D J J, Gernoth K A and Bishop R F 2001 *Phys. Rev. B* **64** 172409
[58] Darradi R, Richter J and Farnell D J J 2005 *Phys. Rev. B* **72** 104425
[59] Darradi R, Derzhko O, Zinke R, Schulenburg J, Krüger S E and Richter J 2008 *Phys. Rev. B* **78** 214415
[60] Bishop R F, Li P H Y, Darradi R and Richter J 2008 *Europhys. Lett.* **83** 47004
[61] Bishop R F, Li P H Y, Darradi R, Richter J and Campbell C E 2008 *J. Phys.: Condens. Matter* **20** 415213
[62] Bishop R F, Li P H Y, Farnell D J J and Campbell C E 2009 *Phys. Rev. B* **79** 174405
[63] Bishop R F, Li P H Y, Farnell D J J and Campbell C E 2010 *Phys. Rev. B* **82** 024416
[64] Bishop R F, Li P H Y, Farnell D J J and Campbell C E 2010 *Phys. Rev. B* **82** 104406
[65] Bishop R F and Li P H Y 2011 *Eur. Phys. J. B* **81** 37

[66] Li P H Y and Bishop R F 2012 *Eur. Phys. J. B* **85** 25
[67] Li P H Y, Bishop R F, Campbell C E, Farnell D J J, Götze O and Richter J 2012 *Phys. Rev. B* **86** 214403
[68] Weihong Z, Oitmaa J and Hamer C J 1991 *Phys. Rev. B* **44** 11869–11881

Journal of Physics: Conference Series **702** (2016) 012002 doi:10.1088/1742-6596/702/1/012002

Phase transition to the Fulde–Ferrell–Larkin–Ovchinnikov state in a quasi-one-dimensional organic superconductor with anion order

Nobumi Miyawaki[1], **Hiroshi Shimahara**[2]

[1] Graduate School of Integrated Arts and Sciences, Hiroshima University, Kagamiyama 1-7-1, Higashi-Hiroshima 739-8521, Japan

[2] Department of Quantum Matter Science, ADSM, Hiroshima University, Kagamiyama 1-3-1, Higashi-Hiroshima 739-8530, Japan

E-mail: nmiyawaki@hiroshima-u.ac.jp

Abstract. A theoretical study is presented on the effect of anion order on the phase transition from the normal state to a Fulde–Ferrell–Larkin–Ovchinnikov superconducting state in a quasi-one-dimensional organic superconductor $(TMTSF)_2ClO_4$. The temperature dependence of the upper critical field $H_{c2}(T)$ is examined. In the absence of anion order, the $H_{c2}(T)$ curve exhibits a one- to two-dimensional crossover with decreasing temperature T. For a sufficiently large anion order parameter, the dimensional crossover disappears and then a kink appears in the $H_{c2}(T)$ curve.

1. Introduction

The realization of the Fulde–Ferrell–Larkin–Ovchinnikov (FFLO) state has been suggested in low-dimensional organic superconductors [1, 2]. The low-dimensionality is advantageous to the FFLO state because the upper critical field H_{c2} is not so strongly reduced by the orbital pair-breaking effect when a magnetic field is applied parallel to the conductive layer. Moreover, anisotropy of the Fermi surface helps stabilize the FFLO state [3, 4].

Recently, Yonezawa *et al.* reported measurements of the superconducting onset temperature $T_c^{\rm onset}$ of a quasi-one-dimensional (Q1D) organic superconductor $(TMTSF)_2ClO_4$ in magnetic fields $\boldsymbol{H}$ aligned with the conductive plane. They found a characteristic field-direction dependence of $T_c^{\rm onset}$ suggestive of the transition to the FFLO superconducting phase [5]. Lebed showed that the observed H_{c2} in $\boldsymbol{H}$ parallel to the b' axis (parallel to the conducting ab plane and perpendicular to the most conductive a axis) agrees in magnitude with a theory in which the FFLO state with d-wave pairing symmetry is assumed [6], Croitoru *et al.* demonstrated how the orbital magnetism in the FFLO state causes the in-plane anisotropy of $T_c^{\rm onset}$ [7]. Miyawaki and Shimahara examined the temperature dependence of H_{c2} with a focus on the effect of the anisotropic Fermi surface and revealed that a novel dimensional crossover of the $H_{c2}(T)$ curve appears for $\boldsymbol{q} \parallel \boldsymbol{a}$ [8], where $\boldsymbol{q}$ is the center-of-mass momentum of Cooper pairs and $\boldsymbol{a}$ is the lattice vector of the most conducting chain of $(TMTSF)_2ClO_4$.

In this paper, the previous theory [8] is extended to investigate $H_{c2}(T)$ in $(TMTSF)_2ClO_4$ on the basis of a more realistic band structure. Specifically, the effect of anion (ClO_4^-) order [9], which is known to cause a splitting of the Q1D Fermi surface, is taken into account. It is shown that the finite anion gap Δ_{AO} gives rise to a qualitative difference in the temperature dependence of the FFLO upper critical field H_{c2}.

2. Model and method

We consider the Q1D tight-binding Hamiltonian [10]

$$H_{AO} = -\sum_{i,j,\sigma} t_{ij} a_{i\sigma}^{\dagger} a_{j\sigma} + \sum_{i,\sigma} \epsilon_i a_{i\sigma}^{\dagger} a_{i\sigma} - \sum_{i,\sigma} (\mu + \sigma\mu_e H) a_{i\sigma}^{\dagger} a_{i\sigma}, \tag{1}$$

where μ is the chemical potential and $\sigma\mu_e H$ ($\sigma = \pm$) is the Zeeman energy. In this model, two kinds of chains (A and B) are defined in the a direction. There is an energy difference between the A- and B-site due to the anion order. This energy difference is accounted for by the parameter

$$\epsilon_i = \begin{cases} \Delta_{AO} & \text{for } i \in A \\ -\Delta_{AO} & \text{for } i \in B \end{cases}. \tag{2}$$

For t_{ij}, only the nearest neighbor hopping is taken into account, and we define $t_{ij} = t_a$ (t_b) for the intra-chain (inter-chain) nearest neighbor hopping.

Diagonalizing this Hamiltonian, we obtain the dispersion [10, 11, 12]

$$\xi_\sigma^s(\boldsymbol{k}, H) = -2t_a \cos k_x - s\sqrt{(2t_b \cos k_y)^2 + {\Delta_{AO}}^2} - \mu - \sigma\mu_e H, \tag{3}$$

where the momentum k_x (k_y) is scaled by the inverse of the lattice constant a (b). There are two electron bands in the Brillouin zone $|k_y| < \pi/2$, corresponding to $s = \pm$. In the organic conductor $(TMTSF)_2ClO_4$, the hole number per site is 0.5 and thus the chemical potential μ is determined by the 1/4-filled condition. The anion gap Δ_{AO} is suggested to be 25 meV by magnetoresistance measurements [11]. Another estimation based on the Hückel method gives $\Delta_{AO} \sim 100$ meV [13]. From recent first-principles calculations, however, Δ_{AO} has been claimed to be nearly zero [14] or $\sim$14 meV [15]. We thus treat Δ_{AO} as a parameter that is less than $t_a \sim 300$ meV.

In the superconducting state, we take into account only intra-band Cooper pairing. We then obtain the following Bogoliubov-quasiparticle energy in an FFLO state with the superconducting gap function $\Delta(\boldsymbol{r}, \boldsymbol{k}) = \Delta(\boldsymbol{k}) e^{i\boldsymbol{q}\cdot\boldsymbol{r}}$:

$$E_{\boldsymbol{k}\sigma}^s = \sigma\zeta^s + E_{\boldsymbol{k}}^s, \tag{4}$$

where

$$E_{\boldsymbol{k}}^s = \sqrt{\xi_\sigma^s(\boldsymbol{k}, 0)^2 + \Delta(\boldsymbol{k})^2}, \tag{5}$$

$$\zeta^s = \frac{1}{2}\boldsymbol{v}_F^s(\boldsymbol{k}) \cdot \boldsymbol{q} - \mu_e H. \tag{6}$$

In Eq. (6), $\boldsymbol{v}_F^s(\boldsymbol{k})$ is the Fermi velocity for the s band.

The gap $\Delta(\boldsymbol{k})$ is determined self-consistently from

$$\Delta(\boldsymbol{k}') = \frac{1}{N} \sum_{s=\pm} \sum_{\boldsymbol{k}} V(\boldsymbol{k}', \boldsymbol{k}) \frac{1 - f(E_{\boldsymbol{k}\uparrow}^s) - f(E_{\boldsymbol{k}\downarrow}^s)}{2E_{\boldsymbol{k}}^s} \Delta(\boldsymbol{k}), \tag{7}$$

where $V(\boldsymbol{k}',\boldsymbol{k}) = V\gamma(\boldsymbol{k}')\gamma(\boldsymbol{k})$ is the pairing interaction. According to Ref. [6], we assume a d-wave symmetry of the gap, i.e., $\Delta(\boldsymbol{k}) = \Delta_0\gamma(\boldsymbol{k})$ with $\gamma(\boldsymbol{k}) = \sqrt{2}\cos k_y$. The normal–superconducting phase boundary is determined by linearizing the above gap equation with respect to Δ_0. From the linearized gap equation, we can obtain

$$\begin{aligned}\ln\frac{T}{T_{\mathrm{c}}^{(0)}} &= -\sum_{s=\pm}\int_{-\pi/2}^{\pi/2}\frac{dk_y}{2\pi}\frac{\rho^s(0,k_y)\gamma^2(k_y)}{N_d(0)}\sinh^2\frac{\beta\zeta^s}{2}\int_0^\infty dy\ln y\\ &\times\left[\frac{2\sinh^2 y}{\left[\cosh^2 y+\sinh^2(\beta\zeta^s/2)\right]^2}-\frac{1}{\cosh^2 y\left[\cosh^2 y+\sinh^2(\beta\zeta^s/2)\right]}\right]\end{aligned} \tag{8}$$

with

$$N_d(0) = \sum_{s=\pm}\int_{-\pi/2}^{\pi/2}\frac{dk_y}{2\pi}\rho^s(0,k_y)\gamma^2(k_y). \tag{9}$$

Here, $\rho^s(0,k_y)$ is the Fermi surface value of the density of states defined through

$$\frac{1}{N}\sum_{\boldsymbol{k}}(\cdots) = \int d\xi\sum_{s=\pm}\int_{-\pi/2}^{\pi/2}\frac{dk_y}{2\pi}\rho^s(\xi,k_y)(\cdots). \tag{10}$$

Equation (8) determines H_{c2} as a function of $(T,\boldsymbol{q})$. An optimum value of $\boldsymbol{q}$ is fixed such that H_{c2} is maximized at a given temperature T, leading to an $H_{\mathrm{c2}}(T)$ phase boundary.

The optimum $\boldsymbol{q}$ occurs when it is directed along $\boldsymbol{a}$, as expected from the Fermi-surface nesting consideration [3, 4, 8, 16]. For the optimum value of $q = |\boldsymbol{q}|$, there are two possibilities associated with the nesting vectors $\boldsymbol{q}^+$ and $\boldsymbol{q}^-$ for the outer $(s=+)$ and inner $(s=-)$ Fermi surfaces (Fig. 1). In fact, when we plot H_{c2} as a function of q, we find that the $H_{\mathrm{c2}}(q)$ curve has two local maxima at low temperatures, though it has a single maximum at high temperatures. We shall denote the critical fields at the two local maxima by $H_{\mathrm{c2}}^{\pm}$.

3. Results

Figure 2 shows plots of the numerical results of $H_{\mathrm{c2}}^{+}(T)$. The FFLO state appears at low temperatures below $T^* \approx 0.56T_{\mathrm{c}}^{(0)}$, where $T_{\mathrm{c}}^{(0)}$ is the transition temperature at $H=0$. When $\varDelta_{\mathrm{AO}} = 0$ (black solid line), $H_{\mathrm{c2}}^{+}(T)$ curve exhibits a characteristic temperature dependence. Just below T^*, the $H_{\mathrm{c2}}^{+}(T)$ curve has a positive curvature as in the case of the 1D system (dashed line); however, as temperature decreases, the curvature becomes negative and a shoulder appears. As temperature decreases further, the curve exhibits an upturn similar to the 2D system (dotted line). This one- to two-dimensional crossover is typical for Q1D systems [8]. As $\varDelta_{\mathrm{AO}}$ is increased, the magnitude of $H_{\mathrm{c2}}^{+}(T)$ decreases, and the crossover behavior becomes less pronounced for $\varDelta_{\mathrm{AO}} \gtrsim 0.3t_a$. This is because the Fermi-surface mismatch between the up- and down-spin Fermi surfaces for $s=-$ at the nesting vector $\boldsymbol{q}^+$ is larger for larger $\varDelta_{\mathrm{AO}}$.

When $\varDelta_{\mathrm{AO}} = 0$, the critical field $H_{\mathrm{c2}}^{+}(T)$ is larger than $H_{\mathrm{c2}}^{-}(T)$ throughout the entire temperature region. However, numerical calculations show that the reduction of $H_{\mathrm{c2}}^{-}(T)$ by a finite $\varDelta_{\mathrm{AO}}$ is less significant than that of $H_{\mathrm{c2}}^{+}(T)$. This is because the density of states $\rho^-(0,k_y)$ is larger than $\rho^+(0,k_y)$ on the nested Fermi surface. As a result, for $\varDelta_{\mathrm{AO}} \gtrsim 0.3t_a$, the critical field $H_{\mathrm{c2}}^{-}(T)$ becomes larger than $H_{\mathrm{c2}}^{+}(T)$ at low temperatures. This means that the upper critical field $H_{\mathrm{c2}}(T)$ observed actually has a kink in its temperature dependence (Fig. 3).

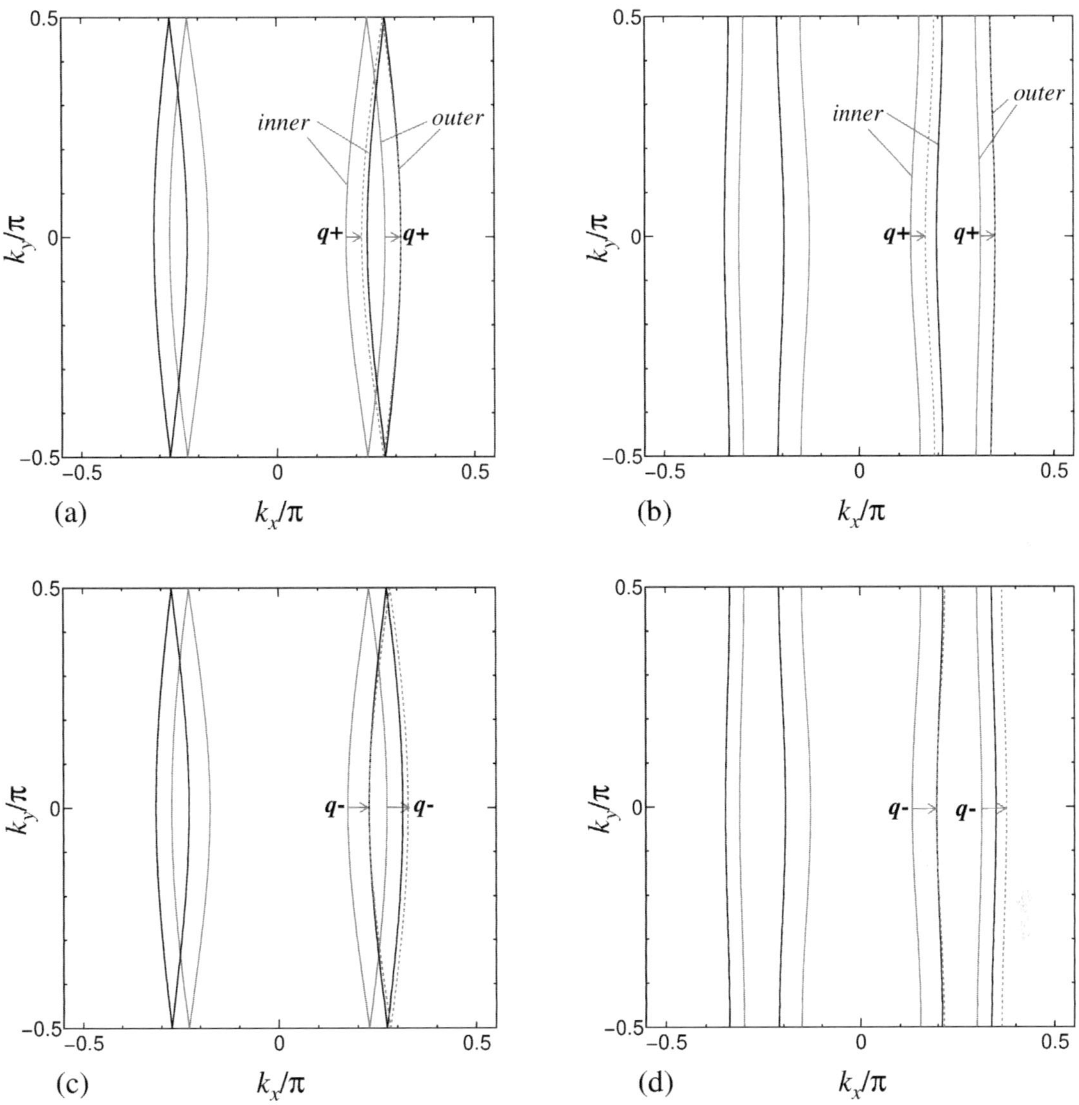

Figure 1. Fermi surfaces for (a),(c) $\Delta_{\rm AO} = 0$ and (b),(d) $\Delta_{\rm AO} = 0.3t_a$. Red and black solid curves are the Fermi surfaces of minority-spin and majority-spin electrons, respectively. The red dotted curves represent the minority-spin Fermi surface shifted by the nesting vector $\boldsymbol{q}^+$ ($\boldsymbol{q}^-$) for the outer (inner) Fermi surface (where the shifted Fermi surface is shown only for the $k_x > 0$ branch).

4. Conclusion

We have examined how the FFLO upper critical field $H_{c2}(T)$ in a Q1D superconductor $(\mathrm{TMTSF})_2\mathrm{ClO}_4$ is modified depending on the anion order. In the absence of the anion gap $\Delta_{\rm AO}$, the $H_{c2}(T)$ curve exhibits a dimensional crossover typical for Q1D systems. For $\Delta_{\rm AO} \gtrsim 0.3t_a$, the dimensional crossover disappears and a kink appears in the $H_{c2}(T)$ curve at a low temperature. If the anion gap $\Delta_{\rm AO}$ is large enough, a kink of $H_{c2}(T)$ should be observed in $(\mathrm{TMTSF})_2\mathrm{ClO}_4$.

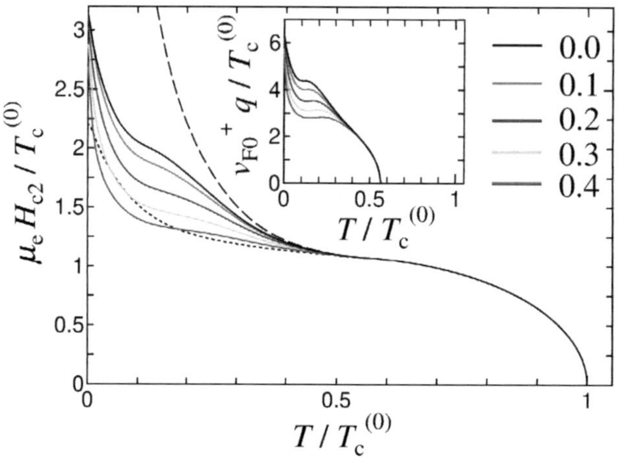

Figure 2. Temperature dependence of $\mu_e H_{c2}^+/T_c^{(0)}$ for $t_b = 0.1t_a$ and $\Delta_{AO} = 0$, $0.1t_a$, $0.2t_a$, $0.3t_a$, and $0.4t_a$. The temperature T is scaled by $T_c^{(0)} = T_c(H = 0)$. For reference, the results for a 1D superconductor ($t_b = 0$ and $\Delta_{AO} = 0$) and for a 2D $d_{x^2-y^2}$-wave superconductor (with a cylindrical Fermi surface) are plotted by dashed and dotted lines, respectively. The inset shows the T dependence of the optimum $v_{F0}^+ q/T_c^{(0)}$. Here, v_{F0}^+ denotes the Fermi velocity $|v_F^+(k_x, k_y = 0)|$.

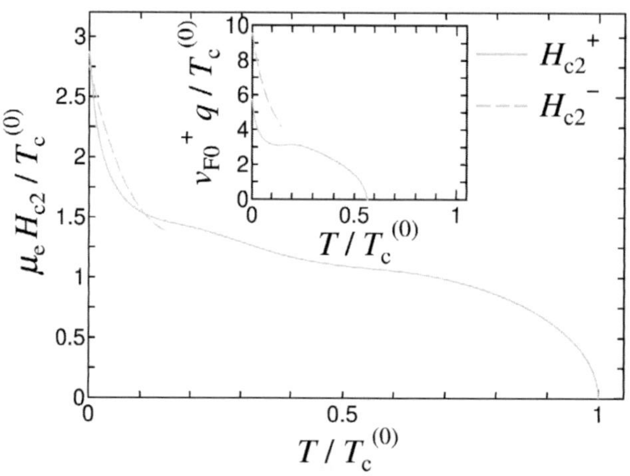

Figure 3. Phase diagram for $t_b = 0.1t_a$ and $\Delta_{AO} = 0.3t_a$. The solid line corresponds to H_{c2}^+ and the dashed line to H_{c2}^-. The inset shows the T dependence of the optimum $v_{F0}^+ q/T_c^{(0)}$. The phase boundary between the normal and superconducting states corresponds to the larger value of H_{c2}^+ and H_{c2}^-. The upper critical fields $H_{c2}^\pm$ cross each other at $T \approx 0.11 T_c^{(0)}$, resulting in a kink in the phase boundary.

Acknowledgments
We would like to thank S. Higashitani for helpful discussions and comments. We also thank S. Yonezawa for useful discussions on experimental data. This work was supported in part by JSPS KAKENHI Grant Number 15K05172.

References

[1] Lebed A G 2008 *The Physics of Organic Superconductors and Conductors* (Berlin: Springer) pp 687–704
[2] Ardavan A, Brown S, Kagoshima S, Kanoda K, Kuroki K, Mori H, Ogata M, Uji S, and Wosnitza J 2012 *J. Phys. Soc. Jpn.* **81** 011004
[3] Shimahara H 1994 *Phys. Rev.* B **50** 12760
[4] Shimahara H 1997 *J. Phys. Soc. Jpn.* **66** 541
[5] Yonezawa S, Kusaba S, Maeno Y, Auban-Senzier P, Pasquier C, Bechgaard K, and Jerome D 2008 *Phys. Rev. Lett.* **100** 117002
[6] Lebed A G 2011 *Phys. Rev. Lett.* **107** 087004
[7] Croitoru M D, Houzet M, and Buzdin A I 2012 *Phys. Rev. Lett.* **108** 207005
[8] Miyawaki N and Shimahara H 2014 *J. Phys. Soc. Jpn.* **83** 024703
[9] At $T = 24$ K, $(TMTSF)_2ClO_4$ shows the orientational order-disorder transition of anion ClO_4^-. Because of the structure of the anion order which is characterized by the wave vector $(0, 1/2, 0)$, the Brillouin zone becomes half. As a result, the Q1D Fremi surface splits at the boundary of the new Brillouin zone.
[10] Shimahara H 2000 *Phys. Rev.* B **61** R14936
[11] Yoshino H, Oda A, Sasaki T, Hanajiri T, Yamada J, Nakatsuji S, Anzai H, and Murata K 1999 *J. Phys. Soc. Jpn.* **68** 3142
[12] Lebed A G, Ha H I, and Naughton M J 2005 *Phys. Rev.* B **71** 132504
[13] Pevelen D L, Gaultier J, Barrans Y, Chasseau D, Castet F, and Ducasse L 2001 *Eur. Phys. J.* B **19** 363
[14] Nagai Y, Nakamura H, and Machida M 2011 *Phys. Rev.* B **83** 104523
[15] Alemany P, Pouget J P, and Canadell E 2014 *Phys. Rev.* B **89** 155124
[16] Miyawaki N and Higashitani S 2015 *Phys. Rev.* B **91** 094511

Journal of Physics: Conference Series **702** (2016) 012003 doi:10.1088/1742-6596/702/1/012003

Charge instabilities of the two-dimensional Hubbard model with attractive nearest neighbour interaction

Raymond Frésard
Laboratoire CRISMAT, UMR 6508 CNRS-ENSICAEN, Caen, France

Kevin Steffen and Thilo Kopp
Center for Electronic Correlations and Magnetism, EP VI, Institute of Physics, University of Augsburg, 86135 Augsburg, Germany

E-mail: Raymond.Fresard@ensicaen.fr

Abstract. Attractive non-local interactions jointly with repulsive local interaction in a microscopic modelling of electronic Fermi liquids generate a competition between an enhancement of the static charge susceptibility—ultimately signalling charge instability and phase separation—and its correlation induced suppression. We analyse this scenario through the investigation of the extended Hubbard model on a two-dimensional square lattice, using the spin rotation invariant slave-boson representation of Kotliar and Ruckenstein. The quasiparticle density of states, the renormalised effective mass and the Landau parameter F_0^s are presented, whereby the positivity of $F_0^s - 1$ constitutes a criterion for stability. Van Hove singularities in the density of states support possible charge instabilities. A (negative) next-nearest neighbour hopping parameter t' shifts their positions and produces a tendency towards charge instability even for low filling whereas the t'-controlled particle-hole asymmetry of the correlation driven effective mass is small. A region of instability on account of the attractive interaction V is identified, either at half filling in the absence of strong electronic correlations or, in the case of large on-site interaction U, at densities far from half filling.

1. Introduction

Attractive nearest neighbour interactions are commonly introduced in electronic lattice models in order to study (unconventional) superconductivity. However, an attractive non-local interaction can also cause a very different phenomenon in the electronic system; in particular, it can generate charge instabilities. In this paper we focus on the tendency towards charge instabilities in two-dimensional (2D) electronic systems which are simultaneously characterised by electronic correlations. Through their correlations the electrons constitute a Fermi liquid, with non-zero Landau parameters and enhanced effective mass. The Fermi liquid behaviour is expected to break down, especially if the correlations are strong close to half filling or when the attractive nearest neighbour interaction dominates.

It is quite conceivable that the on-site interaction of electrons is repulsive whereas the nearest neighbour interaction becomes attractive. Typically, the on-site Coulomb interaction at transition-metal ions is strongly screened by the polarisation of neighbouring atoms, e.g., oxygen ions in cuprates [1, 2]. Yet the screening depends on the local excitations of a cluster of atoms—for example, on the virtual excitations related to charge transfer in CuO_6-octahedra—and can be different for nearest neighbour Coulomb interaction as compared to the local screening or the Lindhard-type screening for longer distances. Coupling to other degrees of freedom [3], such as a strong electron-lattice coupling, may in combination with screening produce an attractive nearest neighbour interaction, as in certain iron-pnictides [4].

In recent years, metallic two-dimensional electron systems were identified at the interface between two insulating films in oxide heterostructures [5, 6]. In contrast to semiconductor interfaces the electronic states in the oxide materials are confined to few atomic layers in the vicinity of the interface (see, e.g. Ref. [7]). Electronic correlation effects have been observed in scanning tunnelling spectroscopy [8], and the electron system was characterised as a 2D electron liquid. Moreover, the compressibility, which is manifestly related to the Landau Fermi-liquid parameter F_0^s, was experimentally observed to be negative in a regime of low charge carrier density [9]—the latter being tuned by a backgate bias.

For the Hubbard model [10], Landau parameters were calculated by Vollhardt [11] within Gutzwiller approximation. Recently, Lhoutellier *et al.* [12] calculated the Landau parameters F_0^s and F_0^a for an extended three-dimensional Hubbard model within a spin rotation invariant generalisation [13, 14] of the Kotliar and Ruckenstein slave-boson (KRSB) representation [15]. For a 2D system the Hubbard model extended by intersite Coulomb interaction and electron-phonon coupling was shown, within a KRSB evaluation, to produce inhomogeneous polaronic states [16], a scenario which may well be realised in some heterostructures. In the present paper we are concerned with a 2D system and we use the same scheme as in Ref. [12] to evaluate the effective mass m^* and F_0^s with focus on attractive non-local interactions and the ensuing charge instabilities. Specifically, we investigate their respective filling dependence. Do electronic correlation effects compete with the tendency towards charge separation?

The KRSB representation was introduced to realise the interaction driven Brinkman-Rice metal-to-insulator transition [17] but since then, it has been successfully used to analyse and characterise antiferromagnetic [18, 19], ferromagnetic [20], spiral [21, 22, 23, 24] and striped [25, 26, 27, 28] phases. KRSB evaluations have been tested against quantum Monte Carlo simulations: A quantitative agreement for charge structure factors was demonstrated [29] and, for example, a very good agreement on the location of the metal-to-insulator transition for the honeycomb lattice has been shown [30]. Also the comparison of ground state energies to numerical solutions [21] or exact diagonalisation data are excellent [23].

The paper is organised as follows: The extended Hubbard model is introduced in Sec. 2, together with its Kotliar and Ruckenstein spin rotation invariant slave-boson representation. In Sec. 3 the saddle-point approximation is presented, jointly with the resulting system of coupled nonlinear equations. Fluctuations are captured within the one-loop approximation, Sec. 4, which allows to determine analytically the Landau parameter F_0^s at half filling for the pure Hubbard model. Numerical results are discussed in Sec. 5, where we address the filling dependence of F_0^s and charge instabilities. The paper is summarised in Sec. 6.

2. Extended Hubbard model

Anderson's suggestion that the simplest model for the d-electrons within the CuO_2 layers common to high-T_c superconductors is the Hubbard model [31] stimulated tremendous research on its properties. However, in the Hubbard model the Coulomb interaction is restricted to the on-site contribution only. In fact, Hubbard himself already argued [10] that for transition metals the matrix elements corresponding to nearest neighbour Coulomb repulsion are relatively large and cannot be disregarded *a priori*. The extended Hubbard model,

$$H = \sum_{i,j,\sigma} t_{ij} c_{i\sigma}^{\dagger} c_{j\sigma} + U \sum_{i} \left(n_{i\uparrow} - \frac{1}{2}\right)\left(n_{i\downarrow} - \frac{1}{2}\right) + \frac{1}{2}\sum_{i,j} V_{ij}(1-n_i)(1-n_j), \quad (1)$$

takes this notion into account by including intersite Coulomb V_{ij} interactions. Although these elements decay fast with increasing distance $|\vec{R}_i - \vec{R}_j|$, they extend in general beyond nearest neighbours. Here $c^\dagger_{i\sigma}$ denotes electron creation operators at site i with spin σ, and $n_{i\sigma} = c^\dagger_{i\sigma} c_{i\sigma}$. The particle-hole symmetric form for both density-density interaction terms is used throughout this work.

In our numerical evaluations below, we restrict the matrix elements t_{ij} to $-t$ for (i, j) a pair of nearest neighbour sites and to $-t'$ for next-nearest neighbour pairs on a square lattice. All other t_{ij} are set to zero. While the bare intersite Coulomb interactions are repulsive, the effective parameters $\{V_{ij}\}$ may become attractive if the coupling to other subsystems is considered [3]. An example is strong electron-lattice coupling, which induces a local lattice deformation surrounding the electron. Under certain conditions such electronic polarons may attract each other as, for example, in some iron-pnictides [4]. Since we are interested in instabilities indicated by $F_0^s \leq -1$, we limit our numerical calculations in this paper to $V_{ij} \leq 0$.

In the spin rotation invariant (SRI) Kotliar and Ruckenstein slave-boson (KRSB) representation [13, 14], which we adopt for our study, one introduces the auxiliary canonical fermionic f_σ, and bosonic e, p_0, $\vec{p}$, and d particles to represent the physical states as:

$$\begin{aligned} |0\rangle &= e^\dagger |\text{vac}\rangle \\ |\sigma\rangle &= \sum_{\sigma'} p^\dagger_{\sigma\sigma'} f^\dagger_{\sigma'} |\text{vac}\rangle \qquad \sigma = \uparrow, \downarrow \\ |2\rangle &= d^\dagger f^\dagger_\uparrow f^\dagger_\downarrow |\text{vac}\rangle , \end{aligned} \tag{2}$$

with $\underline{p}^\dagger = \frac{1}{2}\sum_{\mu=0}^{3} p^\dagger_\mu \underline{\tau}^\mu$, and $\underline{\tau}^\mu$ the Pauli matrices. In terms of these auxiliary operators the Hamiltonian Eq. (1) reads

$$\begin{aligned} H &= \sum_{i,j} t_{ij} \sum_{\sigma\sigma'\sigma_1} z^\dagger_{i\sigma_1\sigma} f^\dagger_{i\sigma} f_{j\sigma'} z_{j\sigma'\sigma_1} + U \sum_i \left(d^\dagger_i d_i - \frac{1}{2}\sum_\sigma f^\dagger_{i\sigma} f_{i\sigma} + \frac{1}{4} \right) \\ &+ \frac{1}{4}\sum_{i,j} V_{ij} \left[\left(1 - \sum_\sigma f^\dagger_{i\sigma} f_{i\sigma}\right) Y_j + Y_i \left(1 - \sum_\sigma f^\dagger_{j\sigma} f_{j\sigma}\right) \right] . \end{aligned} \tag{3}$$

The spin and charge degrees of freedom are mapped onto bosons in this representation; further details are given in Refs. [32] and [12]. We now suppress the site indices in expressions where their reintroduction is self-evident. Above, we introduced the hole doping operator

$$Y \equiv e^\dagger e \; - d^\dagger d \; , \tag{4}$$

while $\underline{z}$ is given by:

$$\underline{z} \equiv e^\dagger \underline{L} \; M \; \underline{R} \; \underline{p} \; + \underline{\tilde{p}}^\dagger \underline{L} \; M \; \underline{R} \; d \tag{5}$$

where $\tilde{p}_{\sigma\sigma'} = (\delta_{\sigma,\sigma'} - \delta_{\sigma,-\sigma'}) p_{-\sigma',-\sigma}$ and

$$\begin{aligned} M &= \left[1 + e^\dagger e + \sum_\mu p^\dagger_\mu p_\mu + d^\dagger d \right]^{\frac{1}{2}} , \\ \underline{L} &= \left[\left(1 - d^\dagger d\right) \underline{1} - 2\underline{p}^\dagger \underline{p} \; \right]^{-\frac{1}{2}} , \\ \underline{R} &= \left[\left(1 - e^\dagger e\right) \underline{1} - 2\underline{\tilde{p}}^\dagger \underline{\tilde{p}} \; \right]^{-\frac{1}{2}} . \end{aligned} \tag{6}$$

The tensor $\underline{z}$ may also be expanded in terms of Pauli matrices as $\underline{z} = \sum_{\mu=0}^{3} z_\mu \underline{\tau}^\mu$. The new Fock space depends in this representation on eight auxiliary operators. The resulting unphysical degrees of freedom may be eliminated through five local constraints,

$$e^\dagger e + \sum_\mu p_\mu^\dagger p_\mu + d^\dagger d \;=\; 1\,, \tag{7}$$

$$\sum_\sigma f_\sigma^\dagger f_\sigma \;=\; \sum_\mu p_\mu^\dagger p_\mu + 2d^\dagger d\,, \tag{8}$$

$$\sum_{\sigma,\sigma'} f_{\sigma'}^\dagger \vec{\tau}_{\sigma\sigma'} f_\sigma \;=\; p_0^\dagger \vec{p} + \vec{p}^\dagger p_0 - i\vec{p}^\dagger \times \vec{p}\ , \tag{9}$$

which are enforced within the path integral formalism and have to be satisfied for each lattice site.

In the SRI KRSB representation the phases of the e and p_μ bosons are gauged away by promoting all constraint parameters to fields [14], and they remain as radial slave-boson fields [33]. Their exact expectation values are generically non-vanishing even though bose condensation is excluded by local gauge invariance [34]. In contrast, the slave-boson field corresponding to double occupancy d is still complex [14, 35, 36].

The saddle-point approximation introduced in the next section is exact in the large degeneracy limit, with Gaussian fluctuations generating the $1/N$ corrections [14], and obeys a variational principle in the limit of large spatial dimensions. In this limit the Gutzwiller approximation becomes exact for the Gutzwiller wave function and longer ranged interactions are static and reduce to their Hartree approximation [37, 38]. On these grounds, the approximation used here to the extended Hubbard model Eq. (1) complies with a variational principle in the limit of large spatial dimensions.

Due to the formal properties mentioned above, our approach covers properties of strongly correlated electrons, as, e.g., the suppression of the quasiparticle residue and the formation of a charge gap at the Mott-Hubbard transition, and the Brinkman-Rice transition [17] to an insulating state at half filling with increasing on-site Coulomb interaction. The impact of the non-local interaction on the latter transition is of particular interest.

3. Saddle-point approximation

Ideally the functional integrals should be calculated exactly. Regarding spin models this has been achieved for the Ising chain [33], but in the case of interacting electron models exact evaluations could be performed on small clusters only, either using the Barnes representation [34], or the Kotliar and Ruckenstein representation [39]. Such a calculation remains challenging on lattices of higher dimensionality, and we here rather resort to the saddle-point approximation. The presentation here for the 2D electronic system follows closely that of Ref. [12] where a 3D extended Hubbard model was investigated. In the translational invariant paramagnetic phase all the local quantities are site independent, and the action at saddle-point reads ($\beta = 1/k_B T$),

$$S = \beta L \left(S_B + S_F + \frac{1}{4} U \right)\,, \tag{10}$$

where L is the number of lattice sites and

$$S_B \;=\; \alpha(e^2 + d^2 + p_0^2 - 1) - \beta_0(p_0^2 + 2d^2) + Ud^2 + \frac{1}{2} V_0 Y\,, \tag{11}$$

$$S_F \;=\; -\frac{1}{\beta} \sum_{\vec{k},\sigma} \ln\left(1 + e^{-\beta E_{\vec{k}\sigma}}\right)\,. \tag{12}$$

Journal of Physics: Conference Series **702** (2016) 012003 doi:10.1088/1742-6596/702/1/012003

Here α and β_0 are site-independent Lagrange multipliers that enforce the constraints Eq. (7) and Eq. (8), respectively. For the extended Hubbard model (1) the quasiparticle dispersion in Eq. (12) reads:

$$E_{\vec{k}\sigma} = z_0^2 t_{\vec{k}} + \beta_0 - \frac{1}{2}U - \frac{1}{2}V_0 Y - \mu \ , \tag{13}$$

Here z_0^2 represents the inverse effective mass m/m^* correction, exclusively caused by correlation effects. The Fourier transform of the intersite Coulomb repulsion is

$$V_{\vec{k}} = \frac{1}{L}\sum_{i,j} V_{ij} e^{-i\vec{k}\cdot(\vec{R}_j - \vec{R}_i)} . \tag{14}$$

It is worth mentioning that only $V_{\vec{k}=0}$ enters Eq. (13). The saddle-point equations following from Eq. (10) read:

$$\begin{aligned} p_0^2 + e^2 + d^2 - 1 &= 0, \\ p_0^2 + 2d^2 &= n, \\ \frac{1}{2e}\frac{\partial z_0^2}{\partial e}\bar{\varepsilon} + \frac{1}{2}V_0\,(1-n)\frac{1}{2e}\frac{\partial Y}{\partial e} &= -\alpha, \\ \frac{1}{2p_0}\frac{\partial z_0^2}{\partial p_0}\bar{\varepsilon} &= \beta_0 - \alpha, \\ \frac{1}{2d}\frac{\partial z_0^2}{\partial d}\bar{\varepsilon} + \frac{1}{2}V_0\,(1-n)\frac{1}{2d}\frac{\partial Y}{\partial d} &= 2\beta_0 - \alpha - U. \end{aligned} \tag{15}$$

Here we have introduced the averaged kinetic energy,

$$\bar{\varepsilon} = \int d\epsilon \rho(\epsilon)\epsilon f_F(z_0^2\epsilon + \beta_0 - \frac{1}{2}U - \frac{1}{2}V_0 Y - \mu) , \tag{16}$$

the determination of which involves the density of states $\rho(\epsilon)$ and $f_F(\dots)$ is the Fermi function. The density of states (DOS) is displayed in Fig. 1(a) for various values of t'. With $y \equiv (e+d)^2$ and the doping away from half filling $\delta = 1 - n$, the equations Eqs. (15) may be merged into a single one:

$$y^3 + (u-1)y^2 = u\delta^2 , \tag{17}$$

where the Coulomb parameter is conveniently expressed in units of U_0 as $u = U/U_0$ with

$$U_0 = -\frac{8}{1-\delta^2}\,\bar{\varepsilon} . \tag{18}$$

The three solutions of Eq. (17) have been discussed in much detail at zero temperature in Ref. [14], with the result that they entail one single critical point located at the Brinkman-Rice transition [17]. It signals a second order transition. At finite temperature, a line of first order transitions with coexistence of phases is found [30]. On the contrary, a finite Hund's coupling in a two-band model induces coexistence of one good and one poor metallic phase at and in the vicinity of half filling at zero temperature [40].

Returning to the one-band model at $\delta = 0$, the inverse effective mass $m/m^* = z_0^2$ may be obtained as

$$z_0^2 = 1 - u^2 . \tag{19}$$

Therefore a metal-to-insulator transition occurs at half filling for a given lattice at U_c, which is defined as follows

$$U_c \equiv \lim_{\delta\to 0} U_0 = -8\bar{\varepsilon} . \tag{20}$$

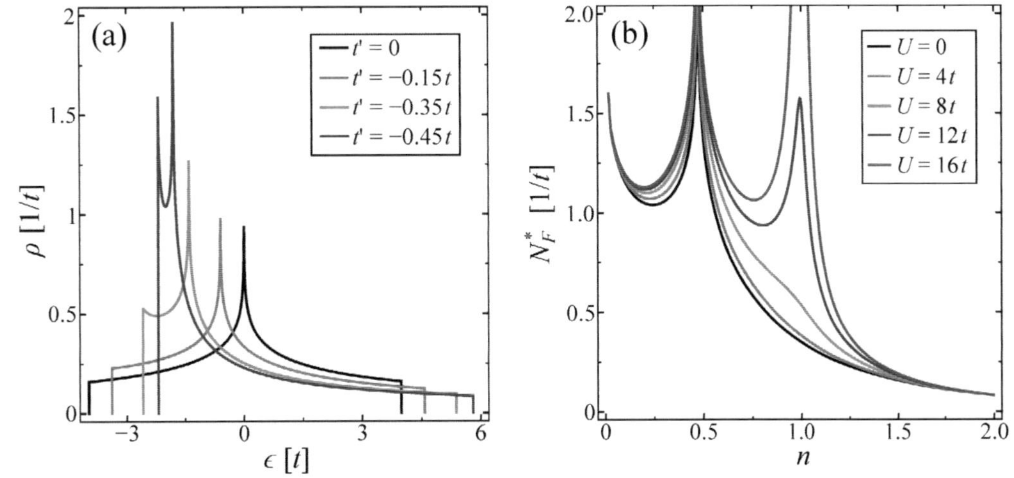

Figure 1. Density of states (DOS). *a*) DOS $\rho(\epsilon)$ of the square lattice for various values of t' as a function of band energy ϵ. *b*) Effective DOS at the Fermi edge N_F^* for $t' = -0.45\,t$ and varying on-site Coulomb interactions U as a function of filling n. The effective DOS is independent of the nearest neighbour interaction V.

A remarkable property of the paramagnetic phase we consider is that V_{ij} elements do not enter explicitly Eq. (17). Hence, in this paramagnetic phase, the intersite interactions only influence the fluctuations and do not change electron localisation due to strong on-site interaction U. In particular, neither a nearest neighbour Coulomb interaction $V_{<i,j>}$ nor a long-ranged one $V(i-j)$ has influence on the Mott gap as discussed by Lavagna [41]. Furthermore, the double occupancy is in exact agreement with the Gutzwiller approximation as derived by Vollhardt, Wölfle and Anderson [42]. In the present case of a 2D square lattice the double occupancy vanishes at half filling for $U_c = 12.96\,t$ with $t' = 0$, $U_c = 13.06\,t$ with $t' = -0.15\,t$, $U_c = 13.4\,t$ with $t' = -0.35\,t$, and for $U_c = 13.62\,t$ with $t' = -0.45\,t$. Thus, the location of the Brinkman-Rice transition [17] shows little dependence on t'. Somewhat surprisingly, the effect of the t'-induced gradual depression of the DOS at half filling and, consequently, the depression of U_c is opposite to the one found for the semi-metallic honeycomb lattice, for which $U_c = 12.6\,t$ was obtained [30].

We here emphasise that Eq. (17) is not only insensitive to V_{ij} elements, but also to the representation of the Hubbard interaction. Indeed, while it is commonly directly expressed in terms of double occupancy, the presently used particle-hole symmetric form results in the same equation. Hence, the energy depends on the precise form of the interactions but the slave-boson saddle-point values do not.

4. One-loop approximation to the charge response function

The mapping of all degrees of freedom onto bosons allows for directly evaluating the charge response function. According to, e. g., Ref. [12], the density fluctuations may be expressed as

$$\delta N \equiv \sum_{\sigma} \delta n_{\sigma} = \delta(d^{\dagger} d - e^{\dagger} e)\,. \tag{21}$$

With this at hand, the charge autocorrelation functions can be written in terms of the slave-boson correlation functions as:

$$\chi_c(k) = \sum_{\sigma\sigma'} \langle \delta n_{\sigma}(-k) \delta n_{\sigma'}(k) \rangle = \langle \delta N(-k) \delta N(k) \rangle. \tag{22}$$

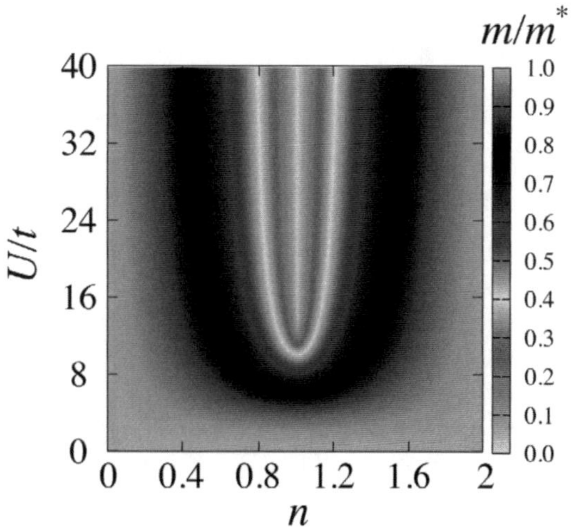

Figure 2. Inverse mass renormalisation for the square lattice with $t' = 0$ as a function of n and U. The effective mass m^* is independent of the nearest neighbour interaction V.

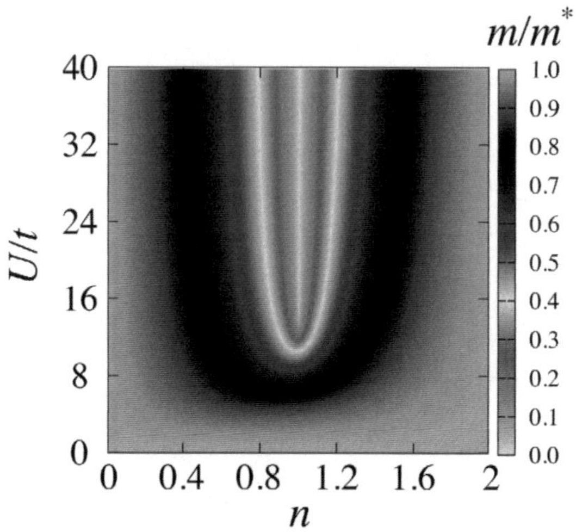

Figure 3. Inverse mass renormalisation for the square lattice with $t' = -0.45\,t$. The effective mass m^* is independent of the nearest neighbour interaction V.

In the following calculation to one-loop order we use the notation $k \equiv (\vec{k}, \omega)$, and the propagator $S_{ij}(k)$ as given in Ref. [12]. The charge susceptibility results as follows:

$$\chi_c(k) = 2e^2 S_{11}^{-1}(k) - 4ed S_{12}^{-1}(k) + 2d^2 S_{22}^{-1}(k). \tag{23}$$

In the long wavelength and low frequency limit, this relation yields the Landau parameter F_0^s. In contrast to the conventional random phase approximation (RPA) results, the obtained Landau parameter cannot be brought to a simple analytical form, unless the intersite Coulomb elements V_{ij} are neglected. In this case, simplifications are most effective at half filling [11, 43, 44], where F_0^s is determined by:

$$F_0^s = -1 + \frac{1}{(1 - U/U_c)^2}\,. \tag{24}$$

The impact of the Hubbard U on F_0^s is transparent: F_0^s steadily increases from 0 at $U = 0$ until it diverges at the metal-to-insulator transition. It should be noted that F_0^s from Eq. (23) is not sensitive to the details of the intersite elements V_{ij}, but only to its zero-momentum Fourier transform (Eq. (14)). For concreteness we restrict ourselves below to nearest neighbour interaction — in which case $V_0 = 4V$ — but our results apply as well to more general situations, provided V is properly interpreted as $V = \frac{1}{4}\frac{1}{L}\sum_{i,j} V_{ij}$.

5. Results

In this work, we evaluated the density response from the one-loop result within slave-boson theory and extracted the dimensionless Landau parameter F_0^s in order to characterise the tendency towards a charge instability through a single effective interaction parameter. Our analysis builds on a 2D lattice model with two interaction scales: A positive on-site Coulomb interaction U, which may be tuned to target the regime of strong electronic correlations, and a negative nearest neighbour interaction V, which can drive a charge instability in the case of $F_0^s \leq -1$.

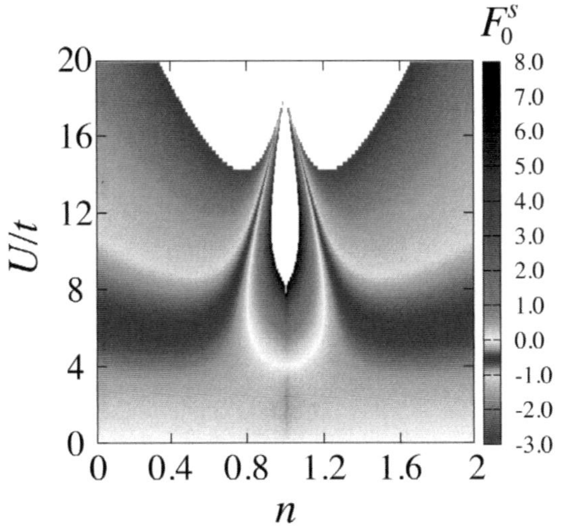

Figure 4. Landau parameter F_0^s for the square lattice as function of n and U. Here $t' = 0$ and $V/U = -0.2$.

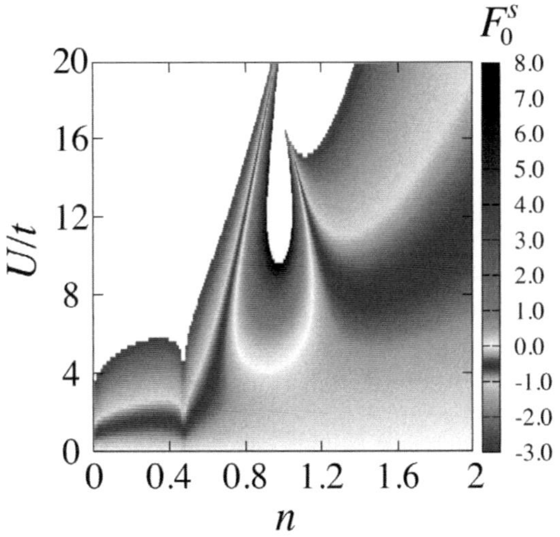

Figure 5. Landau parameter F_0^s for the square lattice. Here $t' = -0.45\,t$ and $V/U = -0.2$.

The question arises if both characteristics appear in this model simultaneously, namely strong correlations and a charge instability—or are they found in separate regimes that are controlled by the electronic density n? Here we may choose to identify strong correlations through a large effective mass m^*. In fact, m/m^* is zero at half filling and small in the respective density regime close to half filling (see Figs. 2 and 3). Moreover, m/m^* is independent of V. Consistent with Fermi liquid theory, the effective density of states at the Fermi energy N_F^* is enhanced with m^*/m, so that one observes three significant structures in $N_F^*(n)$ for finite t' and $U/t \gtrsim 4$ (see Fig. 1(b)): The two lower peaks for $n \to 0$ and $n \simeq 0.5$ represent the Van Hove singularities of the two-dimensional DOS for negative t' (cf. Fig. 1), whereas the buildup of the peak at $n = 1$ is a pure correlation effect. For $t' = 0$ the Van Hove peak and the correlation peak are both placed at $n = 1$, whereas a finite t' serves to keep these peaks apart (which will be helpful also in the discussion of the structure of F_0^s below).

A graphical survey of the impact of positive U and negative V on $F_0^s(n)$ is provided by Figs. 4 and 5. We keep the ratio $V/U = -0.2$ constant in these contour plots in order to reduce the number of control parameters. As required, F_0^s is zero along the $U = 0 = V$ n-axis. For small values of U/t (and correspondingly of $-V/t$), the Landau parameter F_0^s is negative but still larger than -1, irrespective of filling. Then, at intermediate values of U/t, the Landau parameter becomes positive close to half filling, whereas it approaches -1 for filling well below or above half filling. The latter behaviour is induced by the negative nearest neighbour interaction V, whereas the positive value of F_0^s close to half filling is a correlation effect controlled by the repulsive U. Eventually, at large U/t and $-V/t$, Fermi liquid behaviour breaks down, connected either with $m/m^* = 0$ at half filling or with $F_0^s \leq -1$ for finite doping away from half filling.

For $t' < 0$ the (particle-hole) symmetry with respect to $n = 1$ is broken and the charge instability at $F_0^s < -1$ is attained already for lower values of the interaction parameters for $n < 1$ (see Fig. 5). This enhancement of the instability applies especially for values of n at and below the density for which the Van Hove singularity (VHs) produces a peak in the bare $N_F(n)$: A dip structure is formed slightly below $n = 0.5$ (Fig. 5). The VHs is also reflected in the finer dip structure in Fig. 4 at $n = 1$, however there, the correlation-induced increase of F_0^s

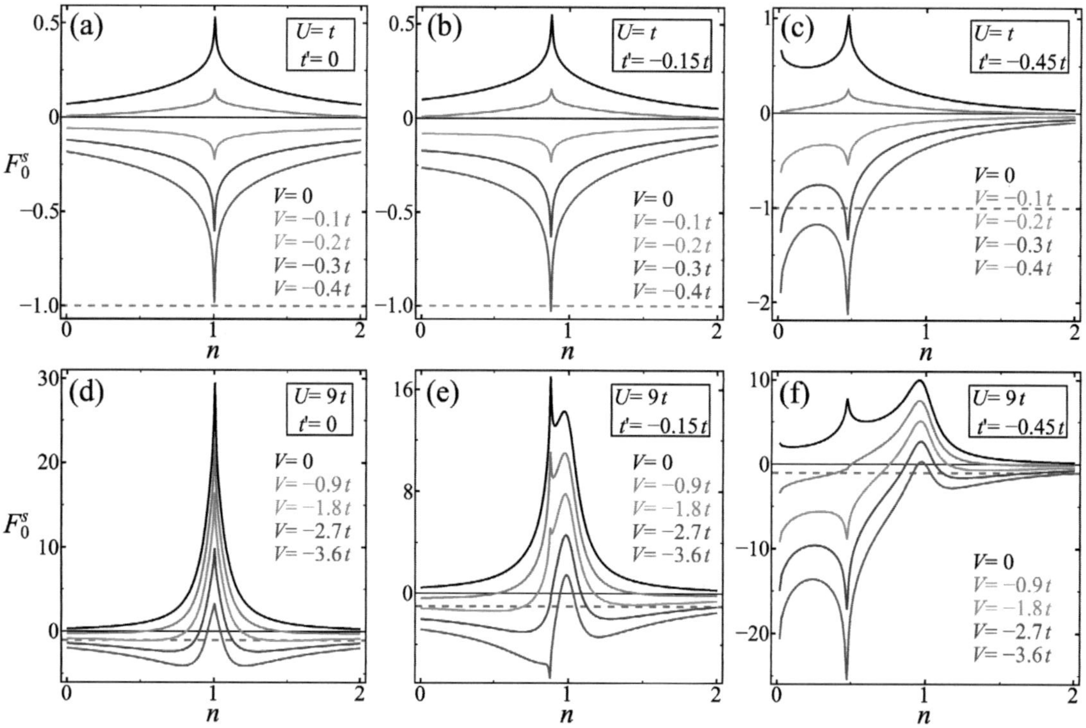

Figure 6. Landau parameter F_0^s for different sets of t' and U as function of filling n. The system becomes instable for $F_0^s < -1$, i.e., below the (brown) dashed line.

and the VHs-controlled decrease of F_0^s through negative V compete close to half filling. This will become evident below, when we fix U and control F_0^s by V.

In Fig. 6 we fix U/t to 1 and 9 in the upper and in the lower set of panels, respectively. In panels (a) and (d) the next-nearest neighbour hopping t' is zero. For $V/t = 0$ and $V/t = -0.1$, F_0^s is positive for all values of n. For sufficiently negative values of V ($V/t \lesssim -0.15$ for $U/t = 1$ in (a)), F_0^s is negative for the full range of n-values. However, for $V/t \gtrsim -0.4$ the electronic system is still stable against charge fluctuations, as $F_0^s > -1$ still applies. For $U/t = 9$ (panel (d)) we observe a peak around $n = 1$ with $F_0^s > 0$. In contrast, F_0^s is negative for n not close to half filling, except for $V = 0$ when $F_0^s \geq 0$ is always true. Obviously, close to half filling Mott-Hubbard-type electronic correlations compete with the charge instability induced through negative V. It depends on the ratio of $U/|V|$ which of them prevails.

For $t' \neq 0$, the picture is more clear-cut (Fig. 6, panel (b) and (e)): For weak coupling, that is $U/t = 1$, the n-dependent structure of F_0^s (peak or dip) is controlled by the position of the VHs (similar to panel (a)). However, for strong coupling $U/t = 9$, a smooth peak emerges around $n = 1$. This peak was concealed in (d) by the contribution from the VHs (at zero or small t') but becomes visible when the VHs-related peak is shifted to lower values of n for more negative values of t'. The shape of the function F_0^s in dependence on n is the central result of our work. At half filling the Mott-Hubbard-type correlations dominate assuming that the negative value of V is not unreasonably large with respect to U.

Below half filling (for negative t'), the position of the VHs determines the peak or dip in $F_0^s(n)$—see also panels (c) and (f). It is not unexpected that $F_0^s(n)$ displays a peak/dip structure controlled by the VHs of the density of states: The dimensionless Landau parameters are composed of a microscopic interaction times the density of states at the Fermi energy.

With this survey of $F_0^s(n)$ in the U-V coupling parameter space, the conclusion can be drawn that charge instabilities induced by an attractive V can only be generated in a regime where strong electronic correlations are absent. The charge instability is boosted by VHs whereas the strong correlations effects are not controlled by VHs.

6. Summary

In order to gain a more thorough understanding of correlated two-dimensional electron systems we applied slave-boson theory to an extended Hubbard model. In the Kotliar and Ruckenstein (spin rotation invariant) slave-boson representation it is feasible to interpolate between the non-interacting limit and the strong coupling case. Here we used this slave-boson representation to determine the two prominent parameters of Landau theory, m^*/m and F_0^s. Beyond the one-band repulsive Hubbard model, we included a nearest neighbour coupling V which is, in our investigation, attractive as we wanted to study instabilities in the (static) charge response. The charge response and, correspondingly, F_0^s were identified in one-loop approximation. While we focused on the effect of nearest neighbour attractive interactions, it should be emphasised that our results for F_0^s also directly apply to more general situations, as the non-local interaction only enters through its zero-momentum Fourier component. Furthermore, we introduced a next-nearest neighbour hopping t' in order to control the position of the Van Hove singularity (VHs)—in particular, to shift the VHs from its position in the middle of the band (at $t' = 0$) to lower energies (for $t' < 0$). Thereby we separated the correlation induced enhancement of F_0^s at half filling from DOS controlled effects.

We found a strong enhancement of m^*/m close to half filling for $U/t \gtrsim 10$, and that mass renormalisation diverges at half filling—in agreement with the standard approaches to the Hubbard model with strong on-site repulsion U. The effective mass m^* turns out to be independent of the nearest neighbour interaction V, at least on the paramagnetic saddle-point level of approximation. Moreover, m^*/m depends only weakly on t'. As expected, the effective mass strongly renormalises the DOS at the Fermi edge (N_F^*) for half filling and intermediate to large values of U/t. When the VHs is shifted towards lower energies for $t' < 0$, then the peak in N_F^* centred at half filling is a signature of the correlation-induced mass enhancement, exclusively.

The Landau parameter F_0^s is zero in the interactionless case. At any finite U and/or V, it is zero only for particular values of filling n, in the case that repulsive U and attractive V compete. Higher Landau parameters may then account for the residual interaction between quasiparticles but this has not yet been investigated.

For weak repulsive on-site interaction $U \simeq t$, the density dependence of F_0^s is dominated by the strong dependence on the VHs which produce peaks of the DOS for distinct fillings. If $-V$ is larger than approximately a tenth of U, the Landau parameter F_0^s attains negative values in the full range of densities. If $|V|$ is increased further, F_0^s becomes less than -1, primarily close to the densities where the VHs dominate, and a charge instability emerges.

However, for intermediate to large values of U/t and $U/|V|$, the charge instability is suppressed close to half filling. In fact, for increasing electronic correlations, we observe the formation of a pronounced peak of F_0^s around half filling. This rounded peak with positive values is the very signature of strong correlations and is entirely absent for weak coupling ($U/t \simeq 1$). The tendency towards charge instability for negative V is quenched by electronic correlations in this regime—below half filling it may reemerge.

Finally, we would like to point to the observation that the effective interaction F_0^s may still be sizable for low densities, both for dominating repulsive U or for attractive V. This is a hallmark of 2D electronic systems, as in contrast F_0^s vanishes in the low density limit for generic 3D lattices, such as the cubic one [12]. This is found both in our formalism and in perturbation theory. In that respect, it is interesting to note that some 2D electronic systems have non-negligible interaction effects even for low electron fillings which are not necessarily related to the presence of long-range Coulomb interactions. Specifically, the electron system at interfaces of $LaAlO_3$ films on $SrTiO_3$ substrates was interpreted as an electron liquid rather than an electron gas (see Ref. [8]). Moreover the metallic state, which is formed on $SrTiO_3$ surfaces appears to support electronic correlations, unexpected for such a low density electron system (see Ref. [45]). Further analyses of these puzzling observations are required in this respect.

Acknowledgements

This work was supported by the DFG through the TRR 80. R.F. is grateful to the Région Basse-Normandie and the Ministère de la Recherche for financial support. The authors acknowledge helpful discussions with V. H. Dao.

7. References

[1] de Boer D, Haas C, and Sawatzky G A, Phys. Rev. B **29**, (1984), 4401.
[2] van den Brink J, *The Hubbard Model with Orbital Degeneracy and in Polarizable Media*, Print Partners Ipskamp (1997).
[3] Micnas R, Ranninger J, and Robaszkiewicz S, Rev. Mod. Phys. B **62** (1990), 113.
[4] Sawatzky G A, Elfimov I S, Van Den Brink J, and Zaanen J, Europhys. Lett. **86** (2009), 17006; Berciu M, Elfimov I, and Sawatzky G A, Phys. Rev. B **79** (2009), 214507.
[5] Ohtomo A and Hwang H Y, Nature **427** (2004) 423.
[6] Thiel S, Hammerl G, Schmehl A, Schneider C W, and Mannhart J, Science **313** (2006), 1942.
[7] Sing M, Berner G, Goß K, Müller A, Ruff A, Wetscherek A, Thiel S, Mannhart J, Pauli S A, Schneider C W, Willmott P R, Gorgoi M, Schäfers F, and Claessen R Phys. Rev. Lett. **102** (2009), 176805.
[8] Breitschaft M, Tinkl V, Pavlenko N, Paetel S, Richter C, Kirtley J R, Liao Y C, Hammerl G, Eyert V, Kopp T, and Mannhart J, Phys. Rev. B **81** (2010) 153414.
[9] Tinkl V, Breitschaft M, Richter C, and Mannhart J, Phys. Rev. B **86** (2012), 075116.
[10] Hubbard J, Proc. R. Soc. London A **276** (1963), 238.
[11] Vollhardt D, Rev. Mod. Phys. **56** (1984), 99.
[12] Lhoutellier G, Frésard R, and Oleś A M, Phys. Rev. B **91** (2015), 224410.
[13] Li T C, Wölfle P, and Hirschfeld P J, Phys. Rev. B **40** (1989), 6817.
[14] Frésard R and Wölfle P, Int. J. of Mod. Phys. B **6** (1992), 685; *ibid.* **6** (1992), 3087.
[15] Kotliar G and Ruckenstein A E, Phys. Rev. Lett. **57** (1986), 1362.
[16] Pavlenko N and Kopp T, Phys. Rev. Lett. **97** (2006), 187001.
[17] Brinkman W F and Rice T M, Phys. Rev. B **2** (1970), 4302.
[18] Lilly L, Muramatsu A, and Hanke W, Phys. Rev. Lett. **65** (1990), 1379.
[19] Yang I, Lange E, and Kotliar G, Phys. Rev. B **61** (2000), 2521.
[20] Möller B, Doll K, and Frésard R, J. Phys.: Condens. Matter **5** (1993), 4847.
[21] Frésard R, Dzierzawa M, and Wölfle P, Europhys. Lett. **15** (1991), 325.
[22] Arrigoni E and Strinati G C, Phys. Rev. B **44** (1991), 7455.
[23] Frésard R and Wölfle P, J. Phys.: Condens. Matter **4** (1992), 3625.
[24] Igoshev P A, Timirgazin M A, Arzhnikov A K, and Irkhin V Y, JETP Lett. **98** (2013), 150.
[25] Seibold G, Sigmund E, and Hizhnyakov V, Phys. Rev. B **57** (1998), 6937.
[26] Seibold G and Lorenzana J, Phys. Rev. B **73** (2006), 144515.
[27] Raczkowski M, Frésard R, and Oleś A M, Phys. Rev. B **73** (2006), 174525; Raczkowski M, Capello M, Poilblanc D, Frésard R, and Oleś A M, *ibid.* **76** (2007), 140505(R).
[28] Raczkowski M, Frésard R, and Oleś A M, Europhys. Lett. **76** (2006), 128.
[29] Zimmermann W, Frésard R, and Wölfle P, Phys. Rev. B **56** (1997), 10097.
[30] Frésard R and Doll K, Proceedings of the NATO ARW *The Hubbard Model: Its Physics and Mathematical Physics*, eds. D. Baeriswyl, D. K. Campbell, J. M. P. Carmelo, F. Guinea, and E. Louis, San Sebastian (1993) (Plenum Press, 1995), p. 385.
[31] Anderson P W, Science **235** (1987), 1196.
[32] Frésard R, Kroha J, and Wölfle P, *Theoretical Methods for Strongly Correlated Systems*, edited by A. Avella and F. Mancini, Springer Series in Solid-State Sciences **171** (Springer-Verlag, Berlin Heidelberg, 2012) pp. 65-101.
[33] Frésard R and Kopp T, Nucl. Phys. B **594** (2001), 769.
[34] Frésard R, Ouerdane H, and Kopp T, Nucl. Phys. B **785** (2007), 286.
[35] Jolicœur Th and Le Guillou J C, Phys. Rev. B **44** (1991), 2403.

[36] Bang Y, Castellani C, Grilli M, Kotliar G, Raimondi R, and Wang Z, Int. J. of Mod. Phys. B **6** (1992), 531; Proceedings of the Adriatico Research Conference and Miniworkshop *Strongly Correlated Electrons Systems III*, eds. Yu Lu, G. Baskaran, A. E. Ruckenstein, E. Tossati (World Scientific Publishing Co., Singapore, 1992).
[37] Metzner W and Vollhardt D, Phys. Rev. Lett. **62** (1989), 324; Phys. Rev. B **37** (1988), 7382; Metzner W, Z. Phys. B **77** (1989), 253.
[38] Müller-Hartmann E, Z. Phys. B **74** (1989), 507.
[39] Frésard R and Kopp T, Ann. Phys. (Berlin) **524** (2012), 175.
[40] Frésard R and Lamboley M, J. Low Temp. Phys. **126** (2002), 1091.
[41] Lavagna M, Phys. Rev. B **41** (1990), 142; Helvetica Phys. Acta **63** (1990), 310; Int. J. Mod. Phys. B **5** (1991), 885.
[42] Vollhardt D, Wölfle P, and Anderson P W, Phys. Rev. B **35** (1987), 6703.
[43] Li T C, Sun Y S and Wölfle P, Z. Phys. B **82** (1991), 369.
[44] Li T C and Bénard P, Phys. Rev. B. **50** (1994), 17837.
[45] Meevasana W, King P D C, He R H, Mo S K, Hashimoto M, Tamai A, Songsiriritthigul P, Baumberger F, and Shen Z X, Nat. Mater. **10** (2011), 114.

Journal of Physics: Conference Series **702** (2016) 012004 doi:10.1088/1742-6596/702/1/012004

On the new Continuous Matrix Product Ansatz

S. S. Chung,[1] **S. Bauman,**[2,1] **Kuei Sun,**[3,1] **and C. J. Bolech**[1]

[1]Department of Physics, University of Cincinnati, Ohio 45221-0011, USA
[2]Dept. of Electrical Eng. and Computer Sys., Univ. of Cincinnati, Ohio 45221-0030, USA
[3]Department of Physics, The University of Texas at Dallas, Texas 75080-3021, USA

Abstract. The fertile new field of quantum information theory is inspiring new ways to study correlated quantum systems by providing fresh insights into the structure of their Hilbert spaces. One of the latest developments in this direction was the extension of the ubiquitous matrix-product-state constructions, epitomized by the density-matrix renormalization-group algorithm, to continuous space-time; so as to be able to describe low-dimensional field theories within a variational approach. Following the earlier success achieved for bosonic theories, we present the first implementation of a continuous matrix product state (cMPS) for spinfull non-relativistic fermions in 1D. We propose a construction of variational matrices with an efficient parametrization that respects the translational symmetry of the problem (without being overly constraining) and readily meets the regularity conditions that arise from removing the ultraviolet divergences in the kinetic energy. We tested the validity of our approach on an interacting spin-1/2 system with spin imbalance. We observe that the ansatz correctly predicts the ground-state magnetic properties for the attractive spin-1/2 Fermi gas, including a phase-oscillating pair correlation function in the partially polarized regime (the 1D correlate of the FFLO state). We shall also discuss how to generalize the cMPS ansatz to other situations.

1. Introduction

Matrix product states (MPS) are a natural class of variational states that arises in several contexts, like in the theory of 1D spin systems [1] or in the framework of the so called Density Matrix Renormalization Group (DMRG) [2, 3, 4]. (DMRG is an optimal variational ansatz for 1D lattice models [5, 6].) A coherent-state extension of the MPS ansatz to the continuum (called cMPS in the literature [7, 8, 9, 10, 11, 12, 13, 14, 15, 16]) has been introduced in the last lustrum. This new ansatz has been very successful but it is still under development. In this proceedings contribution we report on our recent progress in testing and extending cMPS in different directions.

Let us jump ahead and introduce the cMPS directly for the case of fermions. A cMPS for spin-1/2 fermions is the natural generalization of the bosonic case [7] by replacing the boson field by a fermion field and introducing a spin index (σ). Explicitly, for a system length of L and periodic boundary conditions, it has the general form [10]

$$|\chi\rangle = \mathrm{Tr}_{\mathrm{aux}}[\mathcal{P}e^{\int_0^L dx\left[\mathcal{Q}(x)\,\hat{I}+\sum_\sigma \mathcal{R}_\sigma(x)\,\hat{\psi}^\dagger_\sigma(x)\right]}]\,|\Omega\rangle\,, \tag{1}$$

where $\mathcal{Q}(x)$ and $\mathcal{R}_\sigma(x)$ are $\mathbb{C}^{D\times D}$ matrices that act on a D-dimensional auxiliary space and D is called the *bond dimension*. $\hat{I}$ is the identity operator on the Fock space of the bosons ($\hat{\varphi}^\dagger$) or the fermions ($\hat{\psi}^\dagger_\sigma$), $\mathrm{Tr}_{\mathrm{aux}}$ indicates the trace over the auxiliary space, $\mathcal{P}$ is a path-ordering

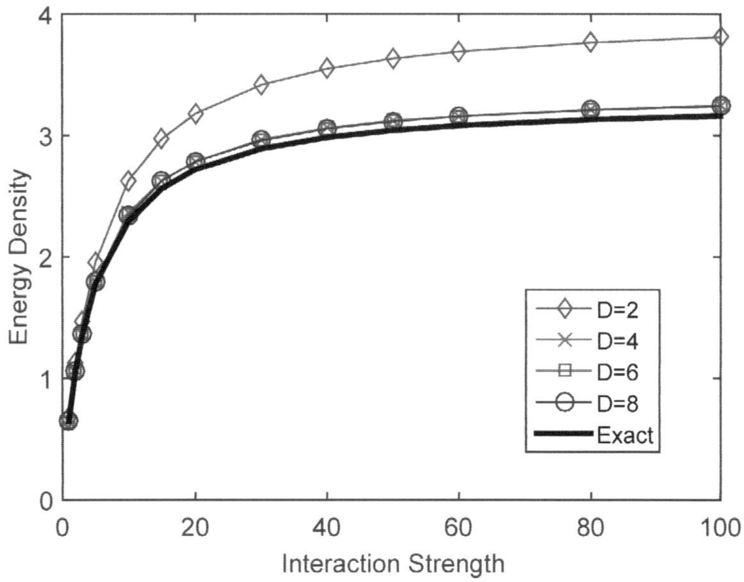

Figure 1. Energy density per particle of the repulsive Lieb-Liniger model as a function of interaction strength (g). The results are given for bond dimensions $D = 2, 4, 6, 8$. The solid line is the exact result as obtained using Bethe's Ansatz.

prescription for the exponential function of Fock-space operators, and $|\Omega\rangle$ is the Fock-space vacuum state.

2. Test Cases

In order to test the performance of the cMPS ansatz and demonstrate its virtues and limitations, we apply it in what follows to three paradigmatic interacting systems, the first two being bosonic and fermionic gases in 1D and the third one a 1D-lattice model of spins. All three models admit exact analytic solutions against which we can compare the cMPS variational results.

2.1. Bosons: the Lieb-Liniger model

The so called Lieb-Liniger model [17] consists of bosons interacting via a repulsive contact potential:

$$H_{\mathrm{LL}} = \int_0^L dx \left(\partial_x \hat{\varphi}^\dagger \partial_x \hat{\varphi} + 2\Delta(\hat{\varphi}^\dagger \hat{\varphi}^\dagger \hat{\varphi}\hat{\varphi})\right), \tag{2}$$

where $\Delta = g/2$ is positive for repulsive interactions.

The results of applying the cMPS variational ansatz to this model were first presented in the same work that introduced the cMPS as an ansatz idea [7]. We show in Fig. 1 how we reproduce those results using our own variational minimization scheme that we will use again below to apply it to other cases. As the bond dimension grows, the cMPS variational results quickly improve and approach the exact result that we obtained via Bethe's Ansatz.

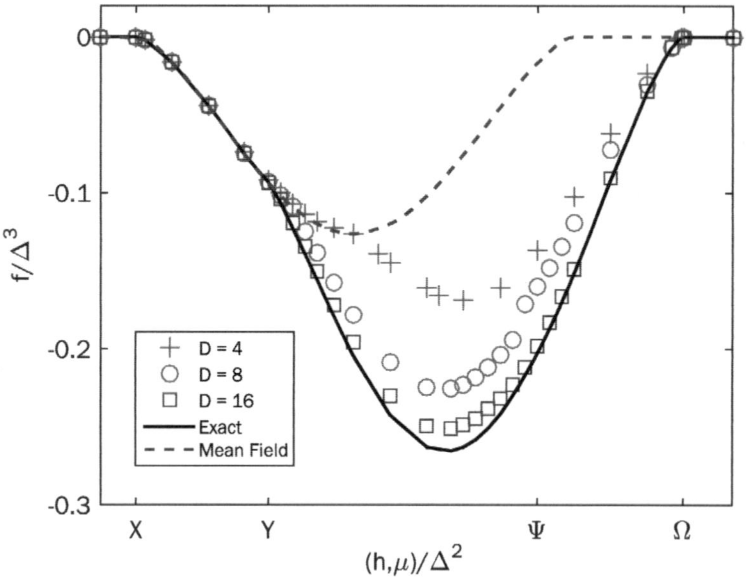

Figure 2. Free energy densities along a circular trajectory centered at the triple point of the attractive Gaudin-Yang model (see Fig. 2 of Ref. [16]). From Ω to Ψ the external magnetic field increases but, due to the competition with pairing due to the interaction, the system remains in a homogeneous quasi-superfluid state. From Ψ to Y the field is even larger and the system remains in a quasi-superfluid state but develops a pair-density-wave order (without breaking translational invariance). Finally, form Y to X the applied field is beyond a density-dependent threshold and the ground state is fully polarized (and thus non-interacting and well described by a mean-field calculation). The results are given for three different bond dimensions, $D = 4, 8, 16$. The solid line is the exact result as obtained using Bethe's Ansatz and the dashed line is the result from a Bogoliubov-deGennes calculation.

2.2. Fermions: the Gaudin-Yang model

The same as for DMRG, which was generalized to fermions (including the calculation of full dynamic correlators [18]) only several years after its initial introduction, the cMPS ansatz was extended to fermions (with full treatment of the interactions) only earlier this year [16]. This was done in the context of interacting 1D gases with spin imbalance that have been recently realized in cold-atom systems [19] and motivated a large amount of work using other techniques [20, 21, 22, 23, 24, 25, 26, 27]. It thus constitutes a highly non-trivial but well understood test case; with the added excitement of being the least controversial concrete realization of the so called FFLO superfluid state [28, 29] (albeit in a one-dimensional sense).

The so called Gaudin-Yang Hamiltonian [30, 31] is the natural generalization of the Lieb-Liniger case to fermions. In this model, spin-1/2 fermions interact via a contact potential and one writes

$$H_{\rm GY} = \int_0^L dx \sum_{\sigma=\uparrow,\downarrow} \left(\partial_x \hat{\psi}^\dagger_\sigma \partial_x \hat{\psi}_\sigma - 2\Delta \hat{\psi}^\dagger_\sigma \hat{\psi}^\dagger_{\bar\sigma} \hat{\psi}_{\bar\sigma} \hat{\psi}_\sigma \right), \tag{3}$$

where $\bar\sigma$ denotes the spin conjugate to σ, and the interaction strength Δ is negative or positive for repulsive or attractive interactions, respectively.

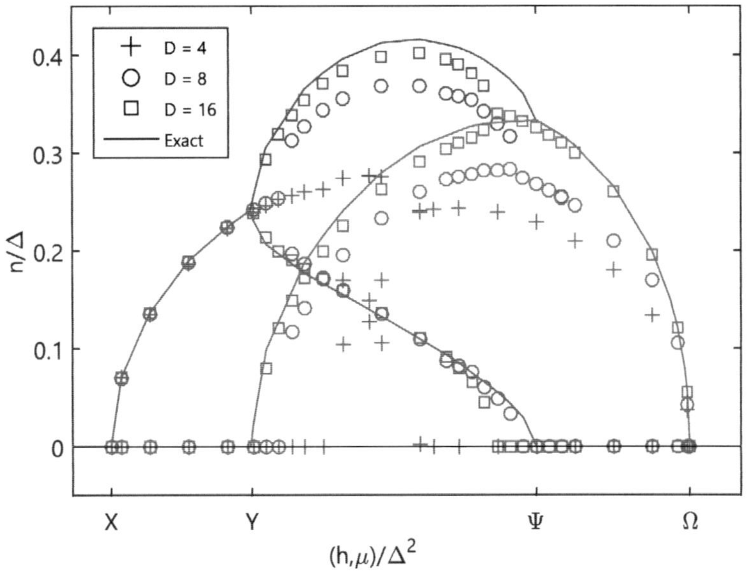

Figure 3. Densities of the majority (blue) and minority (red) fermions, and their difference (green), along the same trajectory as in the previous figure (see Fig. 2). The same three different bond dimensions as before are again compared to the exact results (solid lines).

Applying the fermionic version of the cMPS ansatz we obtained the results shown in Figs. 2 and 3 for the free energy and particle densities, respectively.

Inspired by comparisons with the wave function of Bethe's Ansatz [16], we dropped the spatial dependence of $\mathcal{Q}(x) = Q$ and introduced a phase modulation on $\mathcal{R}_\sigma(x) = R_\sigma e^{iq_\sigma x}$. Now Q and R_σ are complex-valued $D \times D$ matrices independent of position, and the q_σ's are additional real-valued variational parameters.

Regularity of the kinetic-energy density requires imposing the additional conditions:

$$\{R_\uparrow, R_\downarrow\} = 0 \quad \text{and} \quad R_\sigma^2 = 0. \tag{4}$$

which we showed how to implement with great generality in Ref. [16].

In here we introduced an additional overall similarity transformation to the definition of the R_σ matrices, which provides additional freedom to impose the cMPS-gauge condition $Q = -\frac{1}{2}(R_\uparrow^\dagger R_\uparrow + R_\downarrow^\dagger R_\downarrow) - iH$, where H is an arbitrary Hermitian matrix. This additional freedom gives improved free-energy estimates for lower bond dimensions ($D = 4, 8$) and some improvements also for the densities (cf. Ref. [16]), although the determination of the location of some of the phase boundaries is still as delicate as before.

The results are in general good agreement with the exact solution, but cMPS allow us to go further and compute quantities like two- and four-point correlators that are notoriously difficult to obtain from Bethe's Ansatz (this has been shown already in Ref. [16], where we demonstrated the appearance of FFLO-type pair-pair correlations in the relevant part of zero-temperature phase diagram).

The good agreement in the strictly 1D case makes cMPS very promising for extensions to quasi-1D scenarios like the tube arrays that are actually used in the experiments. Other methods have already shown that there is a wide and rich spectrum of physical behaviors to be expected in such systems [32, 33, 34].

Journal of Physics: Conference Series **702** (2016) 012004 doi:10.1088/1742-6596/702/1/012004

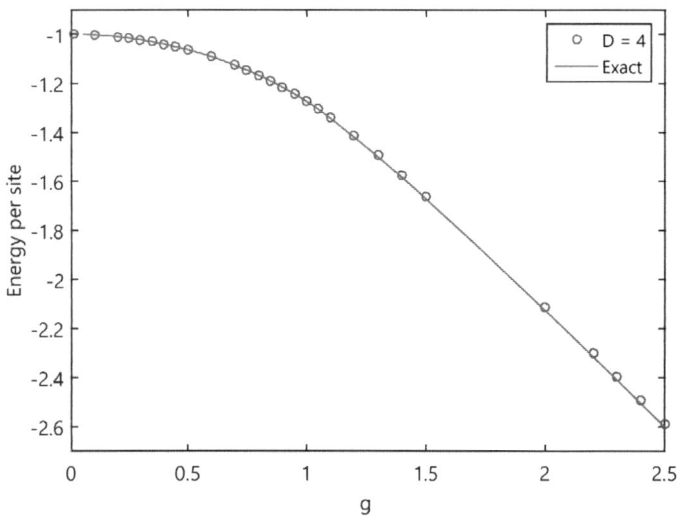

Figure 4. Energy of the transverse-field Ising chain for bond dimension $D = 4$ shown as a function of $g = h/J$. The solid line corresponds to the exact analytic solution of the model.

2.3. Spins: the quantum Ising model

During our presentation, we showed also preliminary results of applying the cMPS ansatz to the quantum Ising model (also known in the literature as the transverse-field Ising model). The Hamiltonian of the system is given by

$$H_{\mathrm{qI}} = -J \sum_{<i,i'>} \sigma_i^x \sigma_{i'}^x - h \sum_i \sigma_i^z, \tag{5}$$

where the angular brackets mean summation over 'near-neighbor pairs'. $J > 0$ corresponds to the ferromagnetic case which we will be studying (many results would be similar for the antiferromagnetic case with $J < 0$). Here $\sigma^{x,z}$ are Pauli matrices and the system is considered in a ring or with periodic boundary conditions (*i.e.*, lattice sites $i = 1$ and $i = L$ are neighbors).

There are a number of ways of applying cMPS to this system, the most direct for us being by mapping first to fermions via a Jordan-Wigner transformation and taking the lattice spacing as if it were the lattice regularization of cMPS. In this way the cMPS ansatz connects naturally to the continuum limit of the model (typically used for analytic studies of its quantum-critical properties).

Results for the energy and the magnetization along the x-axis are shown on Figs. 4 and 5, respectively, and they are compared to the exact solution of the model [35].

While the agreement in terms of energy is uniformly good, the magnetization proves to be a more difficult quantity. The magnetization along the field axis is obtained by direct calculation, with an agreement that is as good as that for the energy (and indeed can also be obtained from the numerical derivative of the result in Fig. 4). But, contrarily, the transverse magnetization is only in good agreement for sub-critical transverse fields. For larger fields cMPS incorrectly gives a non-zero magnetization, while an indirect argument based on the exact solution predicts a sharp transition to a zero value. This is therefore a very sharp transition that provides a stringent test case for cMPS and will be very useful in order to understand it and developing it further (which is work in progress).

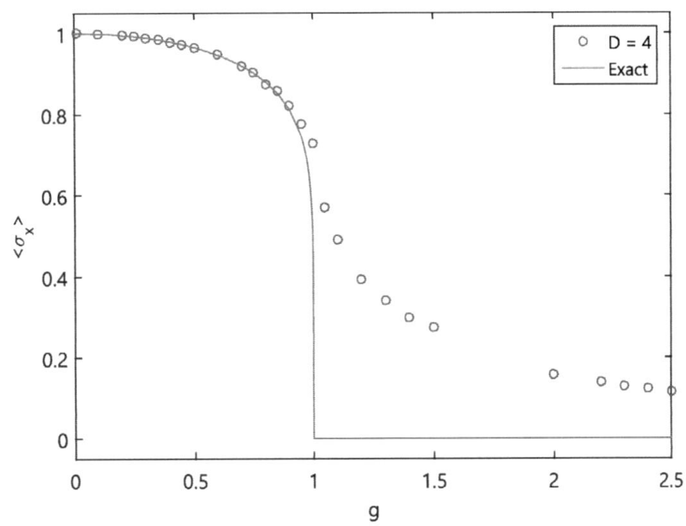

Figure 5. Magnetization along the easy magnetic axis of the exchange interaction, and thus transverse to the field, as a function of $g = h/J$ (cf. Fig. 4). The solid line corresponds to the indirect analytic solution.

3. Conclusions

To summarize, we have demonstrated on three concrete examples the validity and accuracy of cMPS by comparing with available exact solutions. In general, the variational approach can provide additional physical information that is not always easy to extract from the analytic solutions and provides different physical insights. Moreover, as the cMPS ansatz is generic and not restricted to integrable Hamiltonians, we envisage that our work would open the doors to solving outstanding 1D problems and shed light on our understanding of interacting theories of bosons, fermions, spins, or their mixtures.

Acknowledgments

We would like to thank the organizers for giving us the opportunity to present this work during the 18th International Conference on Recent Progress in Many-Body Theories (MBT18) which took place in Niagara Falls, NY USA, in August 2015. We acknowledge federal support from the DARPA OLE program through ARO W911NF-07-1-0464 (SSC, KS, and CJB) for the initial stages of this work. We also received local support by the University of Cincinnati (UC), and in particular SB's work was partially supported by UC's Women in Science & Engineering (WISE) REWU program. Parallel computing resources were from the Ohio Supercomputer Center (OSC) [36].

References

[1] Affleck I, Kennedy T, Lieb E H and Tasaki H 1987 *Phys. Rev. Lett.* **59**(7) 799
[2] Östlund S and Rommer S 1995 *Phys. Rev. Lett.* **75**(19) 3537
[3] Cirac J I and Verstraete F 2009 *J. Phys. A: Math. Theor.* **42**(50) 504004
[4] Schollwöck U 2011 *Annals of Physics* **326**(1) 96
[5] White S R 1992 *Phys. Rev. Lett.* **69**(19) 2863
[6] White S R 1993 *Phys. Rev. B* **48**(14) 10345
[7] Verstraete F and Cirac J I 2010 *Phys. Rev. Lett.* **104**(19) 190405

[8] Maruyama I and Katsura H 2010 *J. Phys. Soc. Jpn.* **79**(7) 073002
[9] Osborne T J, Eisert J and Verstraete F 2010 *Phys. Rev. Lett.* **105**(26) 260401
[10] Haegeman J, Cirac J I, Osborne T J, Verschelde H and Verstraete F 2010 *Phys. Rev. Lett.* **105**(25) 251601
[11] Haegeman J, Cirac J I, Osborne T J and Verstraete F 2013 *Phys. Rev. B* **88**(8) 085118
[12] Draxler D, Haegeman J, Osborne T J, Stojevic V, Vanderstraeten L and Verstraete F 2013 *Phys. Rev. Lett.* **111**(2) 020402
[13] Hübener R, Mari A and Eisert J 2013 *Phys. Rev. Lett.* **110**(4) 040401
[14] Quijandría F, García-Ripoll J J and Zueco D 2014 *Phys. Rev. B* **90**(23) 235142
[15] Stojevic V, Haegeman J, McCulloch I P, Tagliacozzo L and Verstraete F 2015 *Phys. Rev. B* **91**(3) 035120
[16] Chung S S, Sun K and Bolech C J 2015 *Phys. Rev. B* **91**(12) 121108(R)
[17] Lieb E H and Liniger W 1963 *Phys. Rev.* **130**(4) 1605
[18] Kancharla S S and Bolech C J 2001 *Phys. Rev. B* **64**(8) 085119
[19] Liao Y A, Rittner A S C, Paprotta T, Li W, Partridge G B, Hulet R G, Baur S K and Mueller E J 2010 *Nature* **467** 567–569
[20] Yang K 2001 *Phys. Rev. B* **63**(14) 140511
[21] Orso G 2007 *Phys. Rev. Lett.* **98**(7) 070402
[22] Hu H, Liu X J and Drummond P D 2007 *Phys. Rev. Lett.* **98**(7) 070403
[23] Feiguin A E and Heidrich-Meisner F 2007 *Phys. Rev. B* **76**(22) 220508
[24] Casula M, Ceperley D M and Mueller E J 2008 *Phys. Rev. A* **78**(3) 033607
[25] Kakashvili P and Bolech C J 2009 *Phys. Rev. A* **79**(4) 041603
[26] Lu H, Baksmaty L O, Bolech C J and Pu H 2012 *Phys. Rev. Lett.* **108**(22) 225302
[27] Bolech C J, Heidrich-Meisner F, Langer S, McCulloch I P, Orso G and Rigol M 2012 *Phys. Rev. Lett.* **109**(11) 110602
[28] Fulde P and Ferrell R A 1964 *Phys. Rev.* **135**(3A) A550–A563
[29] Larkin A I and Ovchinnikov Y N 1964 *Zh. Eksp. Teor. Fiz.* **47** 1136 [Sov. Phys. JETP **20**, 762 (1965)]
[30] Gaudin M 1967 *Phys. Lett. A* **24**(1) 55
[31] Yang C N 1967 *Phys. Rev. Lett.* **19**(23) 1312
[32] Sun K and Bolech C J 2012 *Phys. Rev. A* **85**(5) 051607(R)
[33] Sun K and Bolech C J 2013 *Phys. Rev. A* **87**(5) 053622
[34] Sun K and Bolech C J 2014 *Phys. Rev. B* **89**(6) 064506
[35] Pfeuty P 1970 *Ann. Phys.* **57**(1) 79
[36] Ohio Supercomputer Center `http://osc.edu/ark:/19495/f5s1ph73`

Signature of the many-particle localization transition in the entanglement spectrum

Richard Berkovits

Department of Physics, Jack and Pearl Resnick Institute, Bar-Ilan University, Ramat-Gan 52900, Israel

E-mail: `richard.berkovits@biu.ac.il`

Abstract. We use the statistical properties of the entanglement spectrum of an interacting disordered system, in order to track the localized to extended transition as function of interaction strength and excitation energy expected from the many-body localization transition. We show that such a transition is indeed observed, although an interesting saturation behavior in the intermediate excitation energy is seen.

1. Introduction

Ideas and measures from the field of quantum information have been recently applied to condensed mater physics [1, 2], mainly in order to identify quantum phase transition (QPT). Entanglement of a many particle system in a pure state, divided into two regions A and B is quantified by different measures (e.g., entanglement entropy, Rényi entropy and entanglement spectrum (ES)) derived from the reduced density matrix of area A, ρ_A or B, ρ_B.

For example, the entanglement entropy measures the von-Neumann (Shannon) entropy of the eigenvalues λ_i^A of ρ_A, such that: $S_A = -\sum_i \lambda_i^A \ln \lambda_i^A$. A different measure is known as the ES corresponding a transformation of the eigenvalues $\varepsilon_i^A = -\ln \lambda_i^A$. Li and Haldane [3] showed that the ES of a partitioned fractional quantum Hall state at $\nu = 5/2$ state resembled the edge excitation spectrum, and thus established a connection between the properties of the ES and the topological order of this state. Expanding on these insights, several authors suggested that the low-energy ES distribution shows some correspondence to the true many-body excitations (MBE) of the partitioned segment (region A), for example the statistical properties of the MBE and ES of disordered systems are the same [4, 5, 6]. The logic behind this proposal is that the reduced density matrix of a region encodes the influence of the connection to the rest of the system on the local MBE of the disconnected region. For low-lying excitations, the phase space is rather low, and therefore the reduced density matrix corresponds rather well to the local MBE spectrum.

Recently, the concept of many-body localization (MBL) [7] has emerged: Under certain conditions the quantum many-body states of the system are localized in the Hilbert (Fock) space resembling the celebrated Anderson localization [8] of a single particle states in a random potential. MBL implies that even after an arbitrary long evolution a state of a decoupled system remains dependent on the initial conditions. It means that time averaging does not result in equipartition distribution and the entropy never reaches its thermodynamic value. In other words the time average is not equivalent to the ensemble average, i.e. the ergodic hypothesis fails.

Numerical studies of many-body models such as the Heisenberg chains of spins $\frac{1}{2}$ in a random field [9] or one-dimensional Josephson arrays [10] provide evidence in favor of the MBL. Recently there appeared a mathematical proof of the existence of MBL transition [11]. It is commonly believed [12] that from the point of view of MBL a generic many-body system behaves similarly to a one-particle problem on the Bethe lattice (random regular graph or disordered Cayley tree [13]). This allows one to develop some intuition about the properties of the system not far from the MBL transition

Here we use the statistical properties of the ES, in order to numerically study the MBL transition. Essentially, we seek a transition from the localized statistics to extended statistics as function of increased interaction strength, or excitation energy. This actually follow in the footsteps of previous studies seeking a signature of the MBL in the MBE spectrum [9, 14, 15], although it has the advantage of being able to address much larger systems due to the use of ES instead of MBE spectrum. It is also connected to recent ideas on the connection between the ES statistics and irreversability [16].

For the disordered single particle spectrum there is an extensive literature on the statistical properties of the energy spectrum in the localized, critical, diffusive and chaotic regimes. Different energy spectrum and wave function statistics, depending on whether the disordered system has time reversal symmetry (Gaussian orthogonal ensemble (GOE)), broken time reversal symmetry (Gaussian unitary statistics (GUE)), spin-orbit interactions (Gaussian symplectic statistics (GSE)) [17, 18, 19]. The energy spectrum statistics can be used to identify the Anderson localization transition [20].

The statistics of MBE in disordered interacting systems have an interesting twist. For non-interacting many-particle systems the level spacing MBE distribution is expected to follow the Poisson distribution for excitation energies above the second spacing, without depending on the single-level distribution [21]. On the other hand, once repulsive interactions between the particles are considered, a transition to the Wigner distribution for higher excitations is observed [22, 23, 24, 25, 9]. This transition may be interpreted as a signature of the MBL transition[9, 14, 15]. One difficulty in studying this transition is that exact diagonalization needed to study excited states is limited to very small systems. Here, the Li and Haldanes' conjecture can come to the rescue, since as we shall demonstrate below, one may extract the ES up to a few tens of states. Thus, a significant number of low lying excitations of a rather large many-body systems are numerically available.

Here we demonstrate that using the ES can actually provide us with some interesting insights into the MBL transition. By monitoring the transition of the ES statistics as function of the interaction strength and level number one can follow the degree to which the excitation is localized or extended. Generally, one expects that the MBL mobility edge will occur at a certain excitation energy corresponding to a level number. As interaction increases the mobility edge should move down to the Fermi energy. Thus, for low-lying levels and weak interactions we expect that the behavior of the ES statistics will follow Poisson, while for higher level numbers or interaction strength it should follow GOE statistics. Thus, by tracking the crossover in the ES statistics we hope to learn about the MBL transition.

2. Model

To study the MBL transition we consider a spinless 1D electrons system of size L with repulsive nearest-neighbor interactions and on-site disordered potential. The Hamiltonian is given by:

$$H = \sum_{j=1}^{L} \epsilon_j \hat{c}_j^\dagger \hat{c}_j - t \sum_{j=1}^{L-1} (\hat{c}_j^\dagger \hat{c}_{j+1} + h.c.) + U \sum_{j=1}^{L-1} (\hat{c}_j^\dagger \hat{c}_j - \frac{1}{2})(\hat{c}_{j+1}^\dagger \hat{c}_{j+1} - \frac{1}{2}), \quad (1)$$

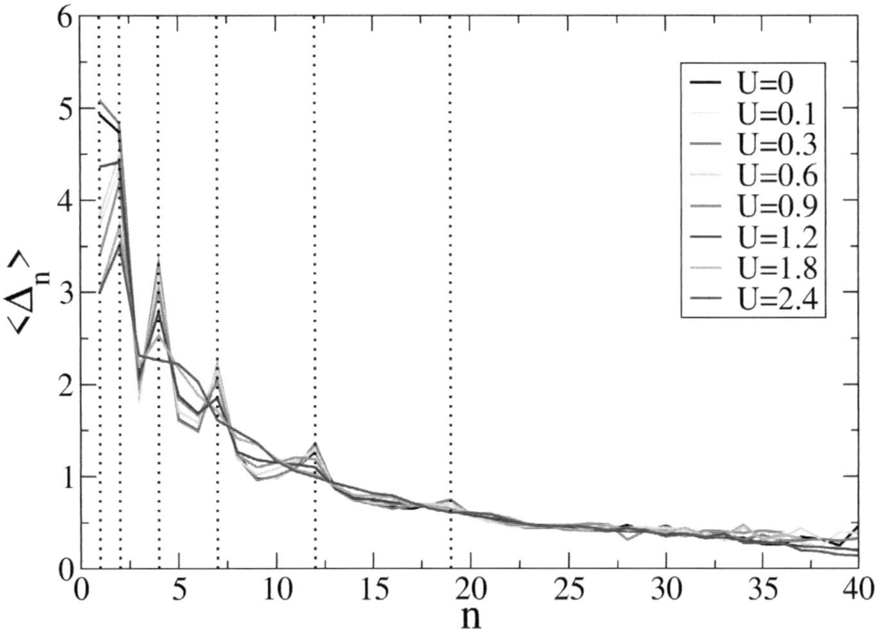

Figure 1. The average ES spacing as function of the level number n for different values of interaction strengths U. The dashed vertical lines correspond to $n = 1, 2, 4, 7, 12, 19$ for which larger spacings are expected [28].

where ϵ_j is the on-site energy drawn from a uniform distribution $[-W/2, W/2]$, $\hat{c}_j^\dagger$ is the creation operator of an electron at site j, and $t = 1$ is the hopping matrix element. The repulsive interaction strength is depicted by $U \geq 0$, and a positive background is considered.

The density matrix renormalization group (DMRG) [26, 27]. is a very accurate numerical method for calculating the ground state of disordered interacting 1D system and for the calculation of the reduced density matrix. We calculate the eigenvalues of ρ_A for a system of length $L = 700$ at half-filling (i.e., the number of electron in the system is $N = L/2 = 350$) and different values of $L_A = 100, 110, \ldots, L - 100$, for at least 20 realizations of disorder. The disorder strength $W = 3.5$ corresponds to a localization length $\xi_0 \approx 105/W^2 \sim 8.5$. The ES is calculated using the relation $\varepsilon_i^A = -\ln \lambda_i^A$.

3. Entanglement spectrum level statistics

The first step in studying the statistical properties of the ES, is to realize that when one bisects the system into regions A and B the number of particles in each region N_A and $N_B = N - N_A$ are still a good quantum number. Since reduced density matrix eigenvalues of different values of N^A do not couple, one can denote the eigenvalues as, $\lambda_n^{N_A}$, which translate into the ES by [3]: $\varepsilon_n^{N_A} = -\ln(\lambda_n^{N_A})$. Thus, for each number sector N_A, the spacing of the ES, $\Delta_n^{N_A} = \varepsilon_{n+1}^{N_A} - \varepsilon_n^{N_A}$, and the average over different lengths L_A, number of particles in the region N_A and realizations depicted by $\langle \Delta_n \rangle$ may be calculated. The results for different strength of interaction U are presented in Fig. 1. Since $\langle \Delta_n \rangle$ should correspond to the many-particle level spacing, it is expected to fall of exponentially as function of n, which is indeed seen. Another interesting feature is the appearance for the non-interacting case of peaks (i.e., larger spacings) for $n = 1, 2, 4, 7, 12, 19$. As discussed elsewhere [28], the origin of these peaks in the

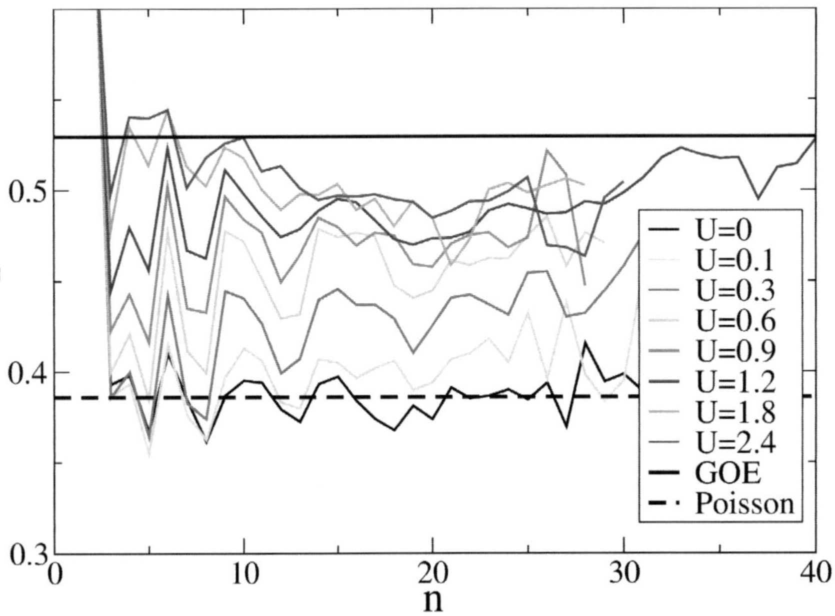

Figure 2. Average r_n as function of level number for different strengths of interaction U. The expected value for r_n in the localized (Poisson) regime is indicated by the heavy dashed black line and in the extended (GOE) regime by a continuous heavy black line.

non-interacting case is essentially combinatoric, where peaks appear when the only possibility of reaching the next excitation energy is by a single-electron transition to a higher energy, contrasted to transitions of several electron, some up some down. These existence of such peaks are an indication of MBL, since they cannot survive the coupling between the many-particle states in the extended states. Indeed, as interaction increase the peaks of Δ_n are erased, and for the highest interactions ($U = 2.4$) only the lowest peaks (at $n = 1, 2$) remain. This is in line with our expectations that as interactions increase the mobility edge goes down, and the localized regime shrinks to the immediate vicinity of the Fermi energy. Nevertheless, this is not an exact method to extract the mobility edge.

In order to get a more quantitative estimation of the localized-extended transition we shall use a statistical measure of the behavior of the spacings between successive ES energies. Following Ref. [15] we define a dimensionless parameter r_n that captures the correlations between successive gaps in the spectrum:

$$\langle r_n \rangle = \frac{\min(\Delta_{\mathrm{n}}^{\mathrm{N_A}}, \Delta_{\mathrm{n+1}}^{\mathrm{N_A}})}{\max(\Delta_{\mathrm{n}}^{\mathrm{N_A}}, \Delta_{\mathrm{n+1}}^{\mathrm{N_A}})}, \tag{2}$$

where $\langle \ldots \rangle$ stands for average over different values of N_A, L_A and realizations of disorder. For the Poisson statistics, $r_n = 2\ln(2) - 1 \sim 0.386$, while for the GOE (Wigner) statistics $r_n \sim .53$. The numerical results of r_n as function of the level number n for different strengths of interaction U are presented in Fig. 2. Let's begin by discussing the non-interacting case $U = 0$. Here we expect that since $\xi \ll L$ all the excited states will show Poisson statistics, i.e., $r_n \sim 0.386$. Indeed, except for the first few values of r_n (especially r_2), which correspond to large level spacings peaks seen in Fig. 1, the values of r_n corresponds very well to the expected GOE value. As U increases, r_n also rises, climbing closer to the GOE value of $r_n \sim .53$. For lower values of interaction ($U < 1$), the larger n is (i.e., the higher the excitation energy) the larger is the increase in r_n. This is in line with the expectation from a mobility edge at higher energy, and with it moving down in energy as interaction increases. At higher interactions ($U > 1$), r_n for higher excitations ($n > 10$) saturates at a value close, but significantly below the GOE value, while r_n for low excitation ($n < 10$) shoots up, reaching the GOE value around $U = 2$. Thus for the highest U we have studied, there is an *intermediate* range of excitations ($10 < n < 30$) for which the statistics of r_n is not quite GOE although it fits very well anywhere else.

4. Discussion

Thus, there is a clear crossover from the Poisson to GOE statistics for the ES statistics in line with the MBL transition. Nevertheless, in order to prove a QPT, we will need to demonstrate finite size behavior corresponding to a mobility edge. There is also the question of the curious saturation of r_n at intermediate excitation numbers. For the ground state of a disordered interacting system interactions drive the system to stronger localization with shorter localization length [29]. On the other hand for the first few excited state, metallic behavior is seen for strong enough interaction [30]. Thus, one naively would expect this metallic behavior to continue for higher excitation, which obviously from Fig. 2 does not occur. This might be connected to the non-ergodic behavior expected in the MBL transition [31], but this needs further study.

Acknowledgments

I would like to thank B. Altshuler, I. Aleiner and V. Oganesyan, for useful discussions.

References

[1] Amico L, Fazio R, Osterloh A and Vedral V 2008 *Rev. Mod. Phys.* **80**, 517
[2] Eisert J, Cramer M, and Plenio M B 2010 *Rev. Mod. Phys.* **82**, 277
[3] Li H and Haldane F D M 2008 *Phys. Rev. Lett.* **101**, 010504
[4] Calabrese P and Lefevre A 2008 *Phys. Rev. A* **78**, 032329
[5] Prodan E, Hughes T L and Bernevig B A 2010 *Phys. Rev. Lett.* **105**, 115501
[6] Gilbert M J, Bernevig B A, and Hughes T L 2012 *Phys. Rev. B* **86**, 041401(R)
[7] Basko D, Aleiner I L and Altshuler B L 2006 *Annals of Physics* **321** 1126
[8] Anderson P W 1958 Phys. Rev. **109**, 651
[9] Oganesyan V and Huse D A 2007 *Phys. Rev. B* **75**, 155111; Oganesyan V, Pal A, and Huse D A 2009 *ibid.* **80**, 115104
[10] Pino M G, Ioffe L B, Altshuler B L arXiv:1501.03853
[11] Imbrie J Z arXiv:math-ph/; Imbrie J Z arXiv:math-ph/1406.2957
[12] Altshuler B L, Gefen Y, Kamenev A, Levitov L 1997 *Phys. Rev. Lett.* **78**, 2803
[13] Bollobás B 2001 *Random Graphs*, (2nd edition, Cambridge University Press).
[14] Pal A and Huse D A 2010 *Phys. Rev. B* **82**, 174411
[15] Iyer S, Oganesyan V, Refael G, and D. A. Huse D A 2013 *Phys. Rev. B*, **87**, 134202
[16] Chamon C, Hamma A, and Mucciolo E R 2014 *Phys. Rev. Lett.* **112**, 240501
[17] Dyson F J and Mehta M L 1963 *J. Math. Phys.* **4**, 701
[18] Gor'kov L P and Eliashberg G M 1965 *JETP* **21**, 940
[19] Altshuler B L and Shklovskii B I 1986 *Zh. Eksp. Teor. Fiz. (Sov. Phys. JETP)* **91** (64), 220 (127)
[20] Shklovskii B I, Shapiro B, Sears B R, Lambrianides P and Shore H B 1993 *Phys. Rev. B* **47**, 11487
[21] Berkovits R 1994 *Europhys. Lett.* **25**, 681
[22] Berkovits R and Avishai Y 1996 *J. of Phys. Condens. Matter* **8**, 389
[23] Pascaud M and Montambaux G 1998 Ann. der Physik **7**, 406 (1998).
[24] Berkovits R and Shklovskii B I 1999 *J. Phys. Condens. Matter* **11**, 779
[25] Song P H, and Shepelyansky D L, *Phys. Rev. B* **61**, 15546
[26] White S R 1992 *Phys. Rev. B* **69**, 2863; 1993 *ibid.* **48**, 10345
[27] Schollwöck U 2005 *Rev. Mod. Phys.* **77**, 259; Hallberg K A 2006 *Adv. Phys.* **55**, 477
[28] Leiman S, Eisenbach A and Berkovits R arXiv:1507.06177
[29] Berkovits R 2012 *Phys. Rev. Lett.* **108**, 176803
[30] Berkovits R 2014 Phys. Rev. B **89** 205137
[31] De Luca A, Altshuler B L, Kravtsov V E, and Scardicchio A 2014 *Phys. Rev. Lett.* **113**, 046806

On the absence of higher generations of incompressible daughter states of composite Fermion quasiparticles

John J. Quinn
University of Tennessee, Knoxville, Tennessee 37996, USA
E-mail: jjquinn@utk.edu

Abstract. Jain [1] introduced a simple mean-field (MF) composite Fermion (CF) picture by attaching to each electron in a quantum Hall system a flux tube producing a Chern- Simons magnetic field $\mathbf{b}(\mathbf{r}) = 2p\phi_0 \sum_i \delta(\mathbf{r} - \mathbf{r}_i)\hat{z}$. Here $\phi_0 = hc/e$ is the quantum of flux, and the sum is over all electron coordinates $\mathbf{r}_i$. He then averaged the total flux and the total charge (electronic plus positive background) over the entire sample. This MF picture gave a system of noninteracting CFs in an effective magnetic field $B_0^* = \nu B_0$. It predicted incompressible quantum liquid (IQL) states at filling factors $\nu = n(1+2pn)^{-1}$ for integral values of n. Chen and Quinn [2] demonstrated that Jain's MF CF picture predicted the total angular momentum values of the lowest energy band of states for any value of the applied magnetic field B_0. Justification of when the MFCF picture was valid was given by Wojs and Quinn [3], who extended the CF hierarchy scheme of Sitko et al. [4]. The CF hierarchy gave the Jain states for integrally filled CF Landau levels (CF LLs), and the Haldane hierarchy of all odd denominator fractions when quasiparticles in the highest (partially filled) CF angular momentum shell had interactions sufficiently similar to the Coulomb interactions of electrons in the lowest Landau level. Sitko et al. showed that the predictions of the CF hierarchy scheme were not always correct. By using a simple pair angular momentum identity and the concept of *fractional grandparentage*, Wojs and Quinn showed that higher generations of CFs could result from the interactions of the original CF quasiparticles only if their interaction energy $V_{\mathrm{QP}}(L_2)$ as a function of their pair angular momentum L_2 increased with increasing L_2 faster than $L_2(L_2+1)$. For Laughlin quasielectrons of the $\nu = 1/3$ IQL state this condition was not satisfied. Therefore, no second generation of CFs could occur. The observed IQL at electron filling factor $\nu = 4/11$ can not be attributed to a daughter IQL state at $\nu_{\mathrm{QE}} = 1/3$ in a totally spin polarized system. Rather, it is an IQL state of electron pairs with pair angular momentum $\ell_{\mathrm{p}} = 2\ell - 1$. A description of the IQL state of electron pairs and its low energy excitations will be presented, and the possibility of a spin flip quantum phase transition as a function of well width and applied magnetic field will be discussed.

1. Introduction

For a system of noninteracting electrons confined to a 2D surface in the presence of a strong perpendicular magnetic field B_0, the density of states, $g(\varepsilon)$, consists of a series of degenerate levels [5] at $\varepsilon_n = \hbar\omega_c(n + 1/2)$ called Landau levels (LLs). For a finite sample of surface area A, the lowest LL can accommodate $(N_\phi + 1)$ electrons, where $N_\phi = B_0 A(hc/e)^{-1}$ is the number of magnetic flux quanta passing through the surface area A. The energy gaps between LLs give rise to the integral quantum Hall (IQH) effect whenever the filling factor $\nu = N(N_\phi + 1)^{-1}$ is an integer [6].

It is convenient to take the 2D surface on which the electrons reside to be a sphere of radius $R = (A/2\pi)^{1/2}$ with a magnetic monopole of strength $2Q$ flux quanta at its center causing a radial magnetic field of magnitude $B_0 = 2Q(hc/e)/(4\pi R^2)$. This spherical geometry [7] has the advantage of a finite surface area with full rotational symmetry. The single particle angular momentum ℓ has a projection ℓ_z satisfying $-\ell \le \ell_z \le \ell$. On the plane $z = 0$, the allowed values of m_z, the z-component of angular momentum, must belong to the set $g_0 = \{0, 1, 2, \cdots, 2\ell\}$. The total angular momentum L of the N particle system has a projection L_z on the sphere. On the plane, M is defined as the sum over all electrons of the value of m_z for each electron. Because ℓ_z and m_z differ by ℓ, it is clear $L_z = M - N\ell$. The eigenstates on the sphere can be written as $|L, L_z\rangle$ and on the plane as $|M_{\rm R}, M_{\rm CM}\rangle$ where $M = M_{\rm R} + M_{\rm CM}$ is the sum of relative and center of mass angular momenta. Interaction energies depend only on L but not L_z, and only on $M_{\rm R}$ but not $M_{\rm CM}$. [8] It is apparent that $M_{\rm R} = N\ell - L$ and $M_{\rm CM} = L + L_z$. To construct an N electron product state of angular momentum $L = 0$, a linear combination of product states with $L_z = 0$ is required, implying that $L = N\ell - M$.

The fractional quantum Hall (FQH) effect discovered by Tsui et al. [9] displayed behavior similar to the IQH effect at specific fractional values of the filling factor ν. The energy gaps giving rise to this FQH effect could only be caused by interactions between the electrons, making it a quantum mechanical many-body problem with no exact solution for $N > 2$. In fact, for $\bar{h}\omega$ much larger than $V_c = e^2/\lambda$, where $\lambda = (\hbar c/eB_0)^{1/2}$ is the magnetic length, the Coulomb scale V_c is the only relevant energy for determining the low energy states. No standard methods based on diagrammatic perturbation theory are valid.

Laughlin [10] suggested that an incompressible quantum liquid (IQL) state occurred whenever it became possible for every pair of electrons to avoid pair states with pair angular momentum, L_2, larger than $2\ell - 3$, the pair state with the largest Coulomb repulsion. The function space of this N-body problem depends on $N_\phi = 2\ell$, which sets the number of single particle functions, and on N, the number of electrons, which must fill a fraction $\nu = N/N_\phi$ of these states. Laughlin suggested a trial wave function $\Psi(1, 2, \cdots, N)$ which was proportional to $\mathcal{F}\{z_{ij}\}\mathcal{G}\{z_{ij}\}$. Here $\mathcal{F} = \prod_{ij} z_{ij}$ is an antisymmetric Fermion factor caused by the Pauli principle, and $\mathcal{G}\{z_{ij}\}$ is a symmetric correlation function caused by the Coulomb interactions. The complex coordinates of the $i^{\rm th}$ electron is denoted by $z_i(= x_i - \imath y_i)$ and $z_{ij} = z_i - z_j$. Laughlin also showed that the elementary excitations of the $\nu = 1/3$ IQL state were quasielectrons (QEs) and quasiholes (QHs) of charge $\mp e/3$. He proposed for the IQL at $\nu = 1/3$ that $\mathcal{G}\{z_{ij}\} = \prod_{i<j} z_{ij}^2$.

Haldane [7] suggested that the problem of putting $N_{\rm QP}$ fractionally charged quasiparticles (QPs) into a QP LL was essentially the same as the original problem of putting N electrons into the electron LL, if the interaction $V_{\rm QP}(L_2)$ between QPs was sufficiently similar to the Coulomb interaction between electrons. This assumption generated a hierarchy of FQH states at all filling factors equal to odd denominator fractions.

2. Composite fermion approach

Jain [1, 11] introduced a composite Fermion (CF) picture by attaching to each electron (via a gauge transformation) a flux tube which carried an even number, $2p$, of magnetic flux quanta. This *Chern-Simons*(CS) flux has no effect on the classical equations of motion, since the Chern-Simons magnetic field $\mathbf{b}(\mathbf{r}) = -2p\phi_0 \sum_i \delta(\mathbf{r} - \mathbf{r}_i)\hat{z}$ vanishes at the position of each electron (it is assumed that no electron senses its own Chern-Simons flux). Here $\phi_0 = hc/e$ is the quantum of flux, and the sum is over all electron coordinates $\mathbf{r}_i$. In a mean-field (MF) approximation, the CS flux and the electron charge are uniformly distributed over the entire area of the sample, giving a system of N composite Fermions (electrons plus CS flux) without interactions (the average charge of $-eN/A$ is canceled by the fixed uniform positive charge background). The effective composite Fermion filling factor ν^* satisfies Jain's equation [1, 11] ${\nu^*}^{-1} = \nu^{-1} - 2p$. When ν^* is an integer $\nu^* = n$, an integral number of CF levels are filled giving an $L = 0$ daughter state at $\nu = n(1 + 2pn)^{-1}$. This Jain sequence of MF CF states (with $n = \pm1, \pm2, \cdots$) is the most robust set of fractional quantum Hall states observed experimentally.

Chen and Quinn [2] introduced an *effective CF angular momentum* ℓ^* satisfying the relation $\ell_0^* = \ell - p(N-1)$, where $2p$ is the number of CS flux quanta attached to each electron to form composite Fermions. Then Jain's simple MF CF picture predicted not only the value of 2ℓ at which an IQL state occurred, but also the total angular momentum values for the lowest band of energy states at any value of the applied magnetic field. This lowest band occurs when the minimum number of QPs consistent with the value of $(2\ell, N)$, defining the function space, is present. The allowed values of the QH and QE angular momenta are $\ell_{\rm QH} = \ell_0^*$ and $\ell_{\rm QE} = \ell_0^* + 1$. The values of the total angular momentum in the lowest band are obtained by angular momentum addition of $n_{\rm QP}$ QP angular momenta $\ell_{\rm QP}$, treating the QPs as identical Fermions. The justification of the simple Jain MF CF picture has been given by Wojs and Quinn [12] and by Benjamin et al. [13] and is reviewed briefly in a later section of this paper.

As an example, consider the 10 electron systems defined by $(2\ell, N) = (25, 10)$ and (29,10). The former case has two QEs, each with $\ell_{\rm QE} = 9/2$. The latter has two QHs, each with $\ell_{\rm QH} = 11/2$. For two identical Fermions, each with angular momentum ℓ, the allowed values of the pair angular momentum are given by $L_2 = 2\ell - j$, where j is an odd integer. This gives [14, 15] the two QE band of $L = 0 \oplus +2 \oplus +4 \oplus +6 \oplus +8$, and the two QH band with $L = 0 \oplus +2 \oplus +4 \oplus +6 \oplus +8 \oplus 10$. In Jain's simple MF picture, the QPs are noninteracting, so that the energies are the same for each value of L_2 ($E = 2\varepsilon_{\rm QE}$ or $E = 2\varepsilon_{\rm QH}$ for the two different cases, where $\varepsilon_{\rm QE}$ and $\varepsilon_{\rm QH}$ are the single QE and QH energies, respectively). Numerical diagonalization studies [14, 15] show that this degeneracy does not occur. In Fig. 1, we reproduce numerical

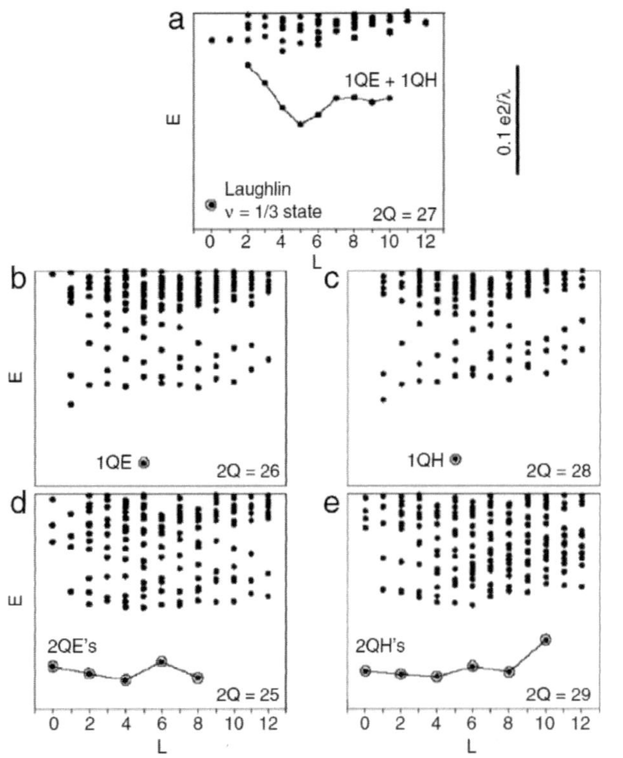

Figure 1. The spectra of 10 electrons in the lowest Landau level calculated on a Haldane sphere with $2Q$ between 25 and 29. The open circles and solid lines mark the lowest energy bands with the fewest composite fermion quasiparticles.(See [15].)

results obtained earlier [15] for an $N = 10$ electron system in the lowest LL (LL0). Frames (d) and (e) contain 2 QEs and 2 QHs, respectively. From the numerical results one can extract $V_{\mathrm{QE}}(L_2)$ and $V_{\mathrm{QH}}(L_2)$, the interaction energies of a pair of QEs (QHs) as a function of the QP pair angular momentum. These interaction energies (or pseudopotentials) are obtained up to an overall constant which has no effect on correlations. In frame (d) there are two QEs each with $\ell_{\mathrm{QE}} = 9/2$, and in frame (d) there are two QHs each with $\ell_{\mathrm{QH}} = 11/2$. The lowest energy bands, separated from a quasi-continuum of higher states by a gap, gives us $V_{\mathrm{QE}}(L_2)$ for $L = 0 \oplus +2 \oplus +4 \oplus +6 \oplus +8$, and $V_{\mathrm{QH}}(L_2)$ for $L = 0 \oplus +2 \oplus +4 \oplus +6 \oplus +8 \oplus 10$. $V_{\mathrm{QE}}(L_2)$ has a maximum at $L_2 = 6$ and minima at $L_2 = 8$ and $L_2 = 4$. $V_{\mathrm{QH}}(L_2)$ has maxima at $L_2 = 10$ and $L_2 = 6$ and minima at $L_2 = 8$ and $L_2 = 4$. This behavior is quite different from the electron pseudopotential in the LL0 which increases monotonically with increasing L_2.

For large systems (e.g. $N > 14$) numerical diagonalization of the electron-electron interactions becomes difficult, so we have investigated the low lying energy states by determining the number of QEs or QHs (n_{QE} or n_{QH}), their angular momenta ℓ_{QE} and ℓ_{QH}, and their interaction energies $V_{\mathrm{QE}}(L_2)$ and $V_{\mathrm{QH}}(L_2)$. Since n_{QE} (or n_{QH}) is much smaller than N, and ℓ_{QE} (and ℓ_{QH}) much smaller than ℓ, the electron angular momentum, we can easily diagonalize these smaller systems. One example [16] is shown in Fig. 2 for the case $(2\ell, N) = (29, 12)$ which corresponds to $(2\ell_{\mathrm{QE}}, n_{\mathrm{QE}}) = (9, 4)$. The low lying states of the electron system are very close to those of the four QE system, suggesting that description in terms of QP excitations interacting via $V_{\mathrm{QP}}(L_2)$ is reasonable.

3. The composite Fermion hierarchy

Sitko et al. [4] introduced a very simple CF hierarchy picture in an attempt to understand Haldane's hierarchy of Laughlin correlated daughter states and Jain's sequence of IQL states with integrally filled CF LLs. Jain's MF CF picture neglected interactions between QPs. The gaps causing incompressibility were energy separations between the filled and lowest empty single particle CF LLs. Not all odd denominator fractions occurred in the Jain sequence $\nu = n(2pn \pm 1)^{-1}$, where n and p are non-negative integers. The missing IQL states were ones with partially filled CF QP shells. The energy gap causing their incompressibility resulted from *residual interactions* between the CF QPs. For an initial electron filling factor ν_0, the

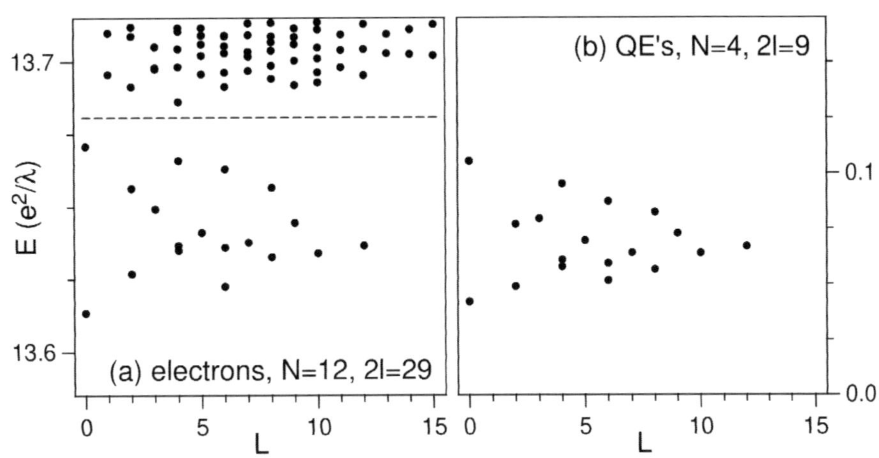

Figure 2. Energy spectra for $N = 12$ electrons in LL0 with $2\ell = 29$, and for $N = 4$ QEs in CF LL1 with $2\ell = 9$. The energy scales are the same, but the QE spectrum was determined using $V_{\mathrm{QE}}(\mathcal{R})$ as the pair pseudopotential (up to an arbitrary constant). (See [16].)

relation between ν_0 and ν_0^*, the effective CF filling factor, satisfied $\nu_0^{-1} = (\nu_0^*)^{-1} + 2p_0$, and gave rise to the Jain states when ν_0^* was equal to an integer n. What happens if ν_0^* is not an integer? It was suggested [4] that then one could write $\nu_0^* = n_1 + \nu_1$, where n_1 was an integer and ν_1 represented the filling factor of the partially filled CF QP shell. If Haldane's assumption that the pair interaction energy $V_{\mathrm{QP}}(L_2)$, as a function of the angular momentum L_2 of the QP pair, was sufficiently similar to $V_0(L_2)$, the interaction energy of the electrons in the LL0, then one could reapply the CF transformation to the CF QPs by writing $(\nu_1^*)^{-1} = \nu_1^{-1} - 2p_1$. Here ν_1 is the CF QP filling factor and $2p_1$ is the number of CS flux quanta added to the original CF QPs to produce a second generation of CFs. For $\nu_1^* = n_2$, an integer, this results in $\nu_1 = n_2(2p_1n_2 \pm 1)^{-1}$, and a daughter IQL state at $\nu_0^{-1} = 2p_1 + \left[n_1 + n_2(2p_1n_2 + 1)^{-1}\right]^{-1}$. This new odd denominator fraction does not belong to the Jain sequence. If ν_1^* is not an integer, then set $\nu_1^* = n_2 + \nu_2$ and reapply the CF transformation to the CF QE in the new QP shell of filling factor ν_2. In general one finds at the ℓth level of the CF hierarchy $\nu_\ell^{-1} = 2p_\ell + (n_{\ell+1} + \nu_{\ell+1})^{-1}$. When $\nu_{\ell+1} = 0$, there is a filled CF shell at the ℓth generation of the CF hierarchy. This procedure generates Haldane's continued fraction leading to IQL states at all odd denominator fractional electron fillings. The Jain sequence is a special case in which $\nu_0^* = n$ gives an integral filling of the first CF QP shell, and the gap is the separation between the last filled and first empty CF levels.

The CF hierarchy picture was tested by Sitko et al. for the simple case of $(2\ell, N) = (18, 8)$ for LL0 by comparing its prediction to the result obtained through exact numerical diagonalization. For this case $2\ell_0^* = 2\ell - 2(N-1) = 18 - 2(7) = 4$. Therefore CF LL0 can accommodate $2\ell_0^* + 1 = 5$ CFs. The three remaining CFs must go into CF LL1 as CF QEs of angular momentum $\ell_{\mathrm{QE}} = \ell_0^* + 1 = 3$. This generates a band of states with $L = 0 \oplus +2 \oplus +3 \oplus +4 \oplus +6$. This is exactly what is found for the lowest energy band of states obtained by numerical diagonalization shown in Fig. 3. Reapplying the CF transformation to the first generation of CF QEs would generate $2\ell_1^* = 2\ell_0^* - 2(N_{\mathrm{QE}} - 1) = 4 - 2(2) = 0$, giving an $L = 0$ daughter IQL state if the CF hierarchy were correct. Clearly the lowest energy state obtained in the numerical diagonalization does not have angular momentum $L = 0$ as predicted by the CF hierarchy. The $L = 0$ and $L = 3$ multiplets clearly have higher energies than the other three multiplets. Sitko et al. conjectured that this must have resulted because the pseudopotential $V_{\mathrm{QE}}(L_2)$ was not sufficiently similar to that of electrons in LL0 to support Laughlin correlations. Laughlin correlations are essential for forming a next generation of CFs.

The QEs and QHs have residual interactions that are more complicated than the simple Coulomb interaction in LL0. We have already seen from Fig. 1(d) and (e), that we can obtain $V_{\mathrm{QP}}(L_2)$ up to an overall constant from numerical diagonalization of N-electron systems in LL0. More careful estimates of $V_{\mathrm{QE}}(\mathcal{R})$ and $V_{\mathrm{QH}}(\mathcal{R})$ (where $\mathcal{R} = 2\ell - L_2$, and L_2 is the pair angular momentum) are shown in Fig. 4. We define a pseudopotential to be *harmonic* if it increases with L_2 as $V_{\mathrm{H}}(L_2) = A + BL_2(L_2+1)$, where A and B are constants. For LL0, the actual pseudopotential $V(L_2)$ always increases with L_2 more rapidly than $V_{\mathrm{H}}(L_2)$. For QEs in CF LL1, the pseudopotential $V_{\mathrm{QE}}(L_2)$ has minima at $L_2 = 2\ell - 1$ and at $L_2 = 2\ell - 5$, and a maximum at $L_2 = 2\ell - 3$. This oscillatory behavior of the interaction energy of a QE pair must be responsible for the failure of the CF hierarchy prediction of an $L = 0$ IQL state.

4. Justification of the CF approach

Pan et al. [17] found IQL states of electrons in LL0 that do not belong to the Jain sequence of integrally filled CF states. One example is the $\nu = 4/11$ filling factor of a state that is assumed to be fully spin polarized. Numerical diagonalization studies of fully spin polarized systems did not find an $L = 0$ IQL ground state at $\nu_{\mathrm{QE}} = 1/3$, which would result in an IQL state at electron filling factor of $\nu = 4/11$. In addition, Pan et al. found strong minima in ρ_{xx} at even denominator filling factors ($\nu = 3/8$ and $\nu = 3/10$) suggesting the existence of IQL states that can't be part of the CF hierarchy. Our research group has made an important contribution to

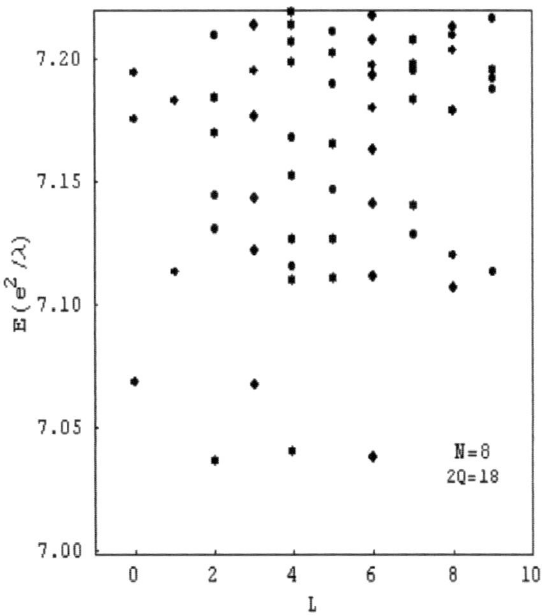

Figure 3. Low energy spectrum of 8 electrons at $2\ell = 18$. The lowest band contains 3 QEs each with $\ell_{\rm QE} = 3$. Reapplying the CS mean-field approximation to these QEs would predict an $L = 0$ daughter state corresponding to $\nu = 4/11$. The data makes it clear that this is not valid. (See [16].)

this field, by rigorously proving [14, 15, 16] under which conditions Jain's elegant CF approach correctly predicts the angular momentum multiplets belonging to the lowest energy sector of the spectrum for any value of the applied magnetic field. Because there is no small parameter in this strongly interacting many body system, our proof does not involve treating fluctuations beyond the MF by a perturbation expansion. It involves proving some rigorous mathematical theorems and applying them, together with well-known concepts frequently used in atomic and nuclear physics. We outline these theorems and give references to earlier publications for proofs. [16]

Theorem 1: Let $\hat{L} = \sum_j \hat{L}_j$ be the total angular momentum operator of an N Fermion system, and $\hat{L}_{ij} = \hat{l}_i + \hat{l}_j$ be the angular momentum operator of the pair $\langle i, j\rangle$. Then

$$\hat{L}^2 + N(N-2)\hat{l}^2 - \sum_{\langle i,j\rangle} \hat{L}_{ij}^2 = 0. \tag{1}$$

where the sum is over all pairs. Here we assume N Fermions are in a shell of angular momentum ℓ, so each Fermion has angular momentum ℓ and projection ℓ_z, with $-\ell \le \ell_z \le \ell$.

Theorem 2: The antisymmetric angular momentum multiplets formed from N Fermions in a shell of angular momentum ℓ can be written as

$$|l^N; L\alpha\rangle = \sum_{L'\alpha'} \sum_{L_2} G_{L\alpha,L'\alpha'}(L_2)|l^2, L_2; l^{N-2}, L'\alpha'; L\rangle \tag{2}$$

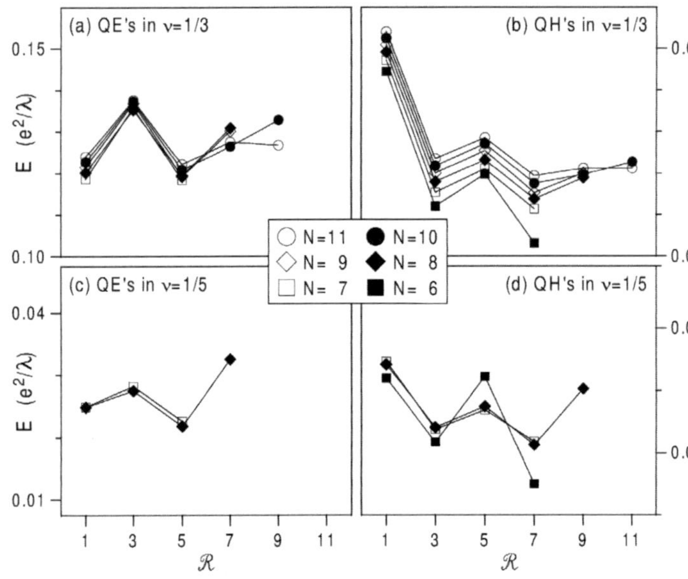

Figure 4. The pseudopotentials of a pair of quasielectrons (left) and quasiholes (right) in Laughlin $\nu = 1/3$ (top) and$\nu = 1/5$ (bottom) states, as a function of relative angular momentum $\mathcal{R}$. Different symbols mark data obtained in the diagonalization of between 6 and 11 electrons.(See [3].)

Here the index α labels different multiplets with the same value of L, and $|l^2, L_2; l^{N-2}, L'\alpha'; L\rangle$ is an N Fermion multiplet of angular momentum L formed from a multiplet $|\ell^{N-2}; L'\alpha'\rangle$ and a pair wave function $|\ell^2, L_2\rangle$. It is antisymmetric with respect to the interchange of indices i and j when both belong to the set (1,2) or when both belong to the set $(3, 4, \cdots, N)$. However, the coefficient $G_{L\alpha,L'\alpha'}(L_2)$, called the coefficient of fractional parentage (CFP) can be chosen so that $|l^N; L\alpha\rangle$ is totally antisymmetric. All that we need to know about the CFP is that

$$\sum_{L'\alpha'} |G_{L\alpha,L'\alpha'}(L_2)|^2 = \mathcal{P}_{L\alpha}(L_2), \tag{3}$$

where $\mathcal{P}_{L\alpha}(L_2)$ is the probability that multiplet $|l^N; L\alpha\rangle$ has pairs with pair angular momentum L_2.

Theorem 3: Taking the expectation value of the operator identity (in Theorem 1) in the state $|l^N; L\alpha\rangle$ *gives*

$$\langle l^N; L\alpha| \sum_{\langle i,j\rangle} \hat{L}_{ij}^2 |l^N; L\alpha\rangle = \frac{1}{2}N(N-1)\sum_{L_2} L_2(L_2+1)\mathcal{P}_{L\alpha}(L_2). \tag{4}$$

This makes use of the fact that $|l^N; L\alpha\rangle$ is totally antisymmetric, so that the sum over all pairs can be replaced by a sum over the allowed values of the pair angular momentum L_2 for any one pair multiplied by the number of pairs, $\frac{1}{2}N(N-1)$.

Theorem 4: If the pseudopotential is harmonic by which we mean $V(L_2) = V_{\rm H}(L_2) = A + BL_2(L_2+1)$ *where* A *and* B *are constants, then every multiplet* α *with the same total angular momentum L has the same energy, given by*

$$E_\alpha(L) = N\left[\frac{1}{2}N(N-1)A + B(N-2)\ell(\ell+1)\right] + BL(L+1). \tag{5}$$

This means that the degeneracy of the angular momentum multiplets of noninteracting Fermions is not removed by a harmonic pseudopotential for different multiplets having the same value of L. [14, 15]

Theorem 5: If $G_{N\ell}(L)$ is the number of independent multiplets of total angular momentum L that can be formed from N Fermions in a shell of angular momentum ℓ, then $G_{N\ell^*}(L) \leq G_{N\ell}(L)$ for every L, if $\ell^* = \ell - (N-1)$. [13]

Theorem 6: The subset $G_{N\ell^*}(L)$ of angular momentum multiplets of the set $G_{N\ell}(L)$ avoids the largest allowed pair angular momentum $L_2 = 2\ell - 1$, which for LL0, corresponds to the largest pair repulsion.

This is obvious for $N = 2$ where $L_2^{\mathrm{MAX}} = 2\ell - 1$ and $L_2^{*\mathrm{MAX}} = 2\ell^* - 1 = 2\ell - 3$, but it is true for arbitrary N. This theorem means that the set of states selected by Jain's MF CF picture (where ℓ^* plays the role of the effective CF angular momentum) is subset of $G_{N\ell}(L)$. This subset avoids pair states with $L_2 = 2\ell - 1$, and contains multiplets with low angular momentum and low energy.

Theorem 7: By adding an integral number, α, of Chern-Simons flux quanta (oriented opposite to the applied magnetic field) to the Hamiltonian for N electrons, not via a gauge transformation, but adiabatically, the pair eigenstate (in the planar geometry) $\Psi_{nm} = e^{im\phi}u_{n,m}(r)$, where $u_{n,m}(r)$ is the radial wave function, transforms to $\tilde{\Psi}_{nm} = e^{im\phi}u_{n,m+\alpha}(r)$.

These theorems justify Jain's CF picture when applied to LL0. Is this important? In our opinion, Jain's MF CF picture is a brilliant success. It is used very often to interpret experimental data. However, because Coulomb and CS gauge interactions beyond the MF involve two entirely different energy scales ($\hbar\omega_c = \nu B$ and $e^2/\lambda \propto \sqrt{B}$), these two interactions between fluctuations beyond MF cannot possibly cancel for all values of B. In addition, treating the interactions by standard many-body perturbation theory can't be rigorously justified because there is no small parameter to assume the convergence of the perturbation series.

Because correlations, that is, the lifting of the degeneracy of the angular momentum multiplets $|l^N; L\alpha\rangle$ of noninteracting electrons in partially filled LL0 depend on the deviation of the actual pseudopotential from the harmonic behavior (i.e. on $\Delta V(L_2) = V(L_2) - V_{\mathrm{H}}(L_2)$), it is interesting to explore the simplest possible anharmonicity. The simplest *anharmonic* contribution to the pseudopotential can be taken as

$$\Delta V(L_2) \equiv V(L_2) - V_{\mathrm{H}}(L_2) = k\ \delta(L_2, 2\ell - 1). \tag{6}$$

If $k > 0$ it is apparent that the lowest energy multiplet for each value of total angular momentum L is the one which avoids (to the maximum possible extent) having pairs with $L_2 = L_2^{\mathrm{MAX}} = 2\ell - 1$. This is exactly what is meant by Laughlin correlations. Complete avoidance of pairs with $L_2 = L_2^{\mathrm{MAX}}$ cannot occur unless $2\ell \geq 3(N-1)$. In the limit of large systems this corresponds to filling factor $\nu \leq 1/3$. If $k < 0$, then the lowest energy state for each value of L is the one with $\mathcal{P}_{L\alpha}(L_2^{\mathrm{MAX}})$ a maximum probability. This corresponds to forming pairs with pair angular momentum $\ell_{\mathrm{p}} = 2\ell - 1$.

It is important to emphasize that Laughlin correlations occur only if $V(\mathcal{R})$ (where $\mathcal{R} = 2\ell - 1$) is *superharmonic* at $\mathcal{R} = 1$. By this we mean that $V(\mathcal{R})$ increases faster than $L_2(L_2+1)$ as L_2 approaches its maximum value of $L_2 = 2\ell - 1$. This is not true of $V_{\mathrm{QE}}(\mathcal{R})$, so we do not expect

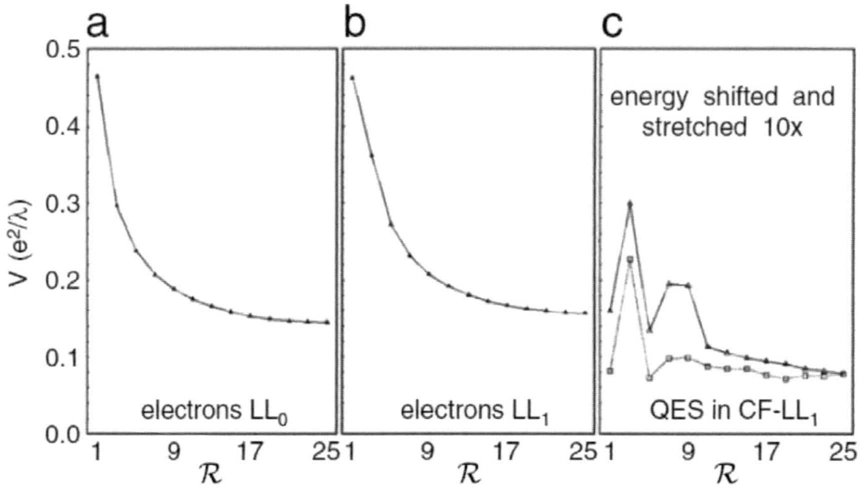

Figure 5. Pair interaction pseudopotentials as a function of relative angular momentum $\mathcal{R}$ for electrons in the LL0 (a), LL1 (b) and for the QEs of the Laughlin $\nu = 1/3$ state by Lee et al.[19] (squares) and by Wojs et al. [20] (triangles). (See [16].)

Laughlin correlations among the CF QEs of the $\nu = 1/3$ Laughlin state at $\nu_{\mathrm{QE}} = 1/3$. This agrees with the numerical result of Sitko et al. [4] Now, however, we understand why the CF hierarchy fails for a spin polarized system. There have been a number of papers [18] suggesting that novel IQL states found by Pan et al. [17] (like the $\nu = 4/11$ state) can be understood as a second generation of CFs forming an $L = 0$ daughter state due to QE interactions. This can't be correct for a spin polarized system with $\nu_{\mathrm{QE}} = 1/3$, because $V_{\mathrm{QE}}(\mathcal{R} = 1)$ will not support Laughlin correlations among the CF QEs.

5. Role of pairing in novel incompressible quantum liquid states

How can the novel IQL states observed by Pan et al. be explained if there is no second generation of CFs formed by Laughlin correlations among the CFs of the first generation? We have shown that if the pseudopotential describing the interaction of a pair of Fermions as a function of the pair angular momentum L_2 is *subharmonic* (i.e. rises with increasing L_2 more slowly than the harmonic pseudopotential V_{H}), the Fermions tend to form pairs with pair angular momentum $L_2 = 2\ell - 1$ instead of being Laughlin correlated. In Fig. 5, we display the pseudopotentials $V^{(0)}(\mathcal{R})$ for electrons in the LL0, $V^{(1)}(\mathcal{R})$ for electrons in LL1, and $V_{\mathrm{QE}}(\mathcal{R})$ for quasielectrons in CF LL1, the first excited CF level.[19, 20] The pseudopotentials are displayed as a function of $\mathcal{R} = 2\ell - L_2$, the *relative pair angular momentum*. $V^{(0)}(\mathcal{R})$ is *superharmonic* at all values of $\mathcal{R}$; $V^{(1)}(\mathcal{R})$ is *superharmonic* for $\mathcal{R} > 3$ but is harmonic or very weakly superharmonic for $\mathcal{R} = 1$. The QE pseudopotential $V_{\mathrm{QE}}(\mathcal{R})$ is not even monotonic; it has strong maxima at $\mathcal{R} = 3$ and 7, and minima at $\mathcal{R} = 1$ and 5.

Moore and Read [21] proposed a *paired wave function* for the observed IQL state at $\nu = 2+\nu_1 = 5/2$, the half-filled state of a spin polarized excited state LL (LL1), sitting above LL0 filled for both spin up and spin down states. The Moore-Read paired wave function (also called the Pfaffian wave function) made use of a correlator based on conformal field theory. Greiter et al. [22] showed that this Phaffian state is an exact solution to a special Hamiltonian, which is large and repulsive when three electrons form single droplets (with total three particle angular momentum $L_3 = 3\ell - 3$ (or relative three particle angular momentum $\mathcal{R}_3 = 3\ell - L_3 = 3$) and zero otherwise. For the Pfaffian wave function at $\nu_1 = 1/2$ in LL1, 2ℓ is given by $2\ell = 2N - 3$ in agreement with numerical diagonalization results. It is worth noting that Laughlin-Jain states in LL0 at filling factors $\nu_1 = n/q < 1/2$ occur when $2\ell = \nu^{-1}N - (q+1-n)$. No even denominator fractional fillings occur in the Laughlin-Jain sequence.

The simplest way of picturing paired states is to introduce a pair angular momentum $\ell_{\rm p} = 2\ell - 1$ and form $N_{\rm p} = N/2$ such pairs. The pairs cannot get too close to one another without violating the Pauli principle. One would normally think of pairs of Fermions as Bosons, but in two dimensional systems we can alter the particle statistics by using a Chern-Simons transformation. We introduce a Fermion pair (FP) angular momentum $\ell_{\rm FP}$ satisfying the equation

$$2\ell_{\rm FP} = 2\ell_{\rm p} - \gamma_{\rm F}(N_{\rm p} - 1). \tag{7}$$

For a single pair $\ell_{\rm FP} = 2\ell - 1$. As $N_{\rm p}$ increases the allowed values of the total angular momentum are restricted to values less than or equal to $2\ell_{\rm FP}$. The value of the constant $\gamma_{\rm F}$ is determined by requiring that the FP filling factor be equal to unity when the single Fermion filling factor has an appropriate value. For $\ell_{\rm p} = 2\ell - 1$, this value corresponds to single Fermion filling $\nu = 1$. Setting $\nu_{\rm FP}^{-1} = (2\ell_{\rm FP} + 1)/N_{\rm p}$, $\nu^{-1} = (2\ell + 1)/N$, and $N_{\rm p} = N/2$ gives $\nu_{\rm FP}^{-1} = 4\nu^{-1} - 3$, (i.e. $\gamma_{\rm F} = 3$), so that $\nu_{\rm FP} = 1$ when $\nu = 1$. The factor of 4 multiplying ν^{-1} results from the pairs having a charge of $-2e$, and $N_{\rm p}$ being equal to $N/2$. This procedure allows the Fermion pairs to be Laughlin correlated instead of the individual electrons being so. It predicts that an IQL state at $\nu_1 = 1/2$ occurs when $2\ell = 2N - 3$ as found by Moore and Read.

We shall apply the same idea to the quasielectrons and quasiholes of the Laughlin $\nu = 1/3$ state, and to quasiholes of the Laughlin-Jain $\nu = 2/5$ state. QHs of $\nu = 1/3$ state reside in CF LL0, but both QEs of the $\nu = 1/3$ state and QHs of the $\nu = 2/5$ state reside in CF LL1.

The pseudopotentials of these quasiparticles are shown in Fig. 6 for a limited range of values of the relative angular momentum $\mathcal{R}$. Note that QEs of the $\nu = 1/3$ state and QHs of the $\nu = 2/5$ state have maxima at $\mathcal{R} = 3$ and minima at $\mathcal{R} = 1$, while QHs of the $\nu = 1/3$ state has maxima at $\mathcal{R} = 1$ and 5 and a minimum at $\mathcal{R} = 3$. The behavior of QHs of the Laughlin $\nu = 1/3$ state leads to avoidance of $\mathcal{R} = 1$ and 5, so we assume the formation of pairs in this case with $\ell_{\rm p} = 2\ell - 3$ instead of $2\ell - 1$.

If we assume that the QEs form pairs and treat the pairs as Fermions, then Eq. (7) gives the relation between the *effective FP angular momentum* $\ell_{\rm FP}$, and the QE angular momentum ℓ, and the relation between the *effective FP filling factor* $\nu_{\rm FP}$, and the QE filling factor $\nu_{\rm QE}$. If

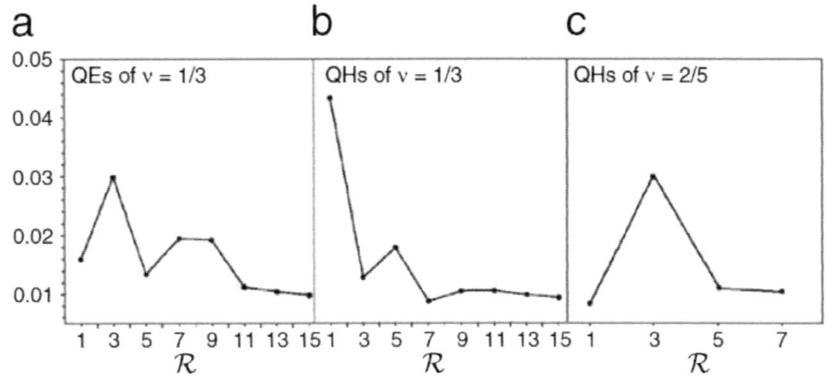

Figure 6. The pseudopotentials $V_{\rm QE}(\mathcal{R})$ and $V_{\rm QH}(\mathcal{R})$ for (a) QEs of $\nu = 1/3$ state, (b) QHs of $\nu = 1/3$, and (c) QHs of $\nu = 2/5$ state. (See [16].)

Table 1. Values of $\nu_{\mathrm{FP}}^{-1} = m$ for $m = 3$, 5, 7, and 9 and the resulting values of ν_{QE}, ν_{QH}, and the electron filling factor that they generate.

ν_{FP}^{-1}	3	5	7	9
ν_{QE}	2/3	1/2	2/5	1/3
ν	5/13	3/8	7/19	4/11
ν_{QH}(CF LL1)	2/3	1/2	2/5	1/3
ν	4/11	3/8	8/21	5/13

Table 2. Values of ν_{QH} satisfying $1/3 > \nu_{\mathrm{QH}} \geq 1/5$ and the resulting electron filling factors ν for Laughlin correlated QP_2s with ν_{FP}^{-1} =7, 9, 11, and 13.

ν_{FP}^{-1}	7	9	11	13
ν_{QH}	2/7	1/4	2/9	1/5
ν	5/17	3/10	7/23	4/13

we take $\nu_{\mathrm{FP}} = m^{-1}$, where m is an odd integer, we can obtain the value of ν_{QE} corresponding to the Laughlin correlated state of FPs (pairs of quasielectrons with $\ell_{\mathrm{p}} = 2\ell - 1$). Exactly the same procedure can be applied to QHs in CF LL1 since $V_{\mathrm{QE}}(\mathcal{R})$ and $V_{\mathrm{QH}}(\mathcal{R})$ are qualitatively similar at small values of $\mathcal{R}$. Here we are assuming that $V_{\mathrm{QE}}(\mathcal{R})$ and $V_{\mathrm{QH}}(\mathcal{R})$ are dominated by their short range behavior $\mathcal{R} \leq 5$. The QH pseudopotential is not as well determined for $\mathcal{R} > 5$ because it requires larger N electron systems than we can treat numerically. The electron filling factor is given by $\nu^{-1} = 2 + (1 + \nu_{\mathrm{QE}})^{-1}$ or by $\nu^{-1} = 2 + (2 - \nu_{\mathrm{QH}})^{-1}$. This results in the values of ν shown in Table 1 for $2/3 \geq \nu_{\mathrm{QE}} \geq 1/3$.

The states generated at the values of ν_{QE} and ν_{QH} equal to 2/5 have not been observed. Clear IQL states were observed by Pan et al.[17] at $\nu = 3/8$ and $\nu = 4/11$. A somehow weaker IQL state at $\nu = 5/13$ is also observed.

The daughter states generated by QPs in CF LL1 (QEs of the parent $\nu = 1/3$ Laughlin state or QHs of the parent $\nu = 2/5$ Jain state) give rise to filling factors for the electron system with $\nu > 1/3$. QHs of the $\nu = 1/3$ Laughlin state (residing in CF LL0) can also form daughter states, and they result in an electron filling factor ν in the range $1/3 > \nu \geq 1/5$. The pseudopotential for these QHs is *superharmonic* at $\mathcal{R} = 1$ and has a strong minimum at $\mathcal{R} = 3$. Because of this, if they form pairs, the pairs must have angular momentum $\ell_{\mathrm{p}} = 2\ell - 3$ (instead of $\ell_{\mathrm{p}} = 2\ell - 1$ for QE pairs). Then, Eq.(7) must be modified. We replace $2\ell - 1$ in the definition of $2\ell_{\mathrm{FP}}$ by $2\ell - 3$, and γ_{F} by $\tilde{\gamma}_{\mathrm{F}}$. The value of $\tilde{\gamma}_{\mathrm{F}}$ is determined by requiring that $\nu_{\mathrm{FP}} = 1$ when $\nu_{\mathrm{QH}} = 1/2$. This condition results from the fact that the pairs are formed by two QHs separated by two filled CF states. The resulting value of $\tilde{\gamma}_{\mathrm{F}}$ is 7, so that $\nu_{\mathrm{FP}}^{-1} = 4\nu_{\mathrm{QE}}^{-1} - 3$ is replaced by $\nu_{\mathrm{FP}}^{-1} = 4\nu_{\mathrm{QH}}^{-1} - 7$. The QH daughter states resulting from Laughlin correlated QH_2 (pairs of QHs of the $\nu = 1/3$ state) and the electron filling factor satisfying $\nu^{-1} = 2 + (1 - \nu_{\mathrm{QH}})^{-1}$ are given in Table 2.

All of the electron filling factors except $\nu = 7/19$ and $\nu = 8/21$ have been observed with minima in ρ_{xx} (or plateaus in ρ_{xy} in many cases). Numerical diagonalization studies seem to support the occurrence of IQL states. However, recent studies by Samkharadze et al.[23] at very low temperatures raise questions about the even denominator fractions.

6. Summary and conclusions

In this paper we have reviewed exact numerical diagonalization of small systems within the Hilbert subspace of a single partially occupied LL. The numerical results are thought of as *numerical experiments*, and simple intuitive models fitting the numerical data are sought, to better understand the underlying correlations. We describe calculations for N electrons confined to a Haldane spherical surface, and present simple results at different values of the LL degeneracy $g = 2\ell + 1$. We demonstrate that Jain's remarkable CF picture predicts not only the values of 2ℓ at which incompressible quantum liquid ground states occur for different values of N, but also predicts the angular momenta L of the lowest band of multiplets for any value of 2ℓ in a very simple way. We emphasize that Jain's CF picture is valid, not because of some magical cancellations of Coulomb and Chern-Simons gauge interactions beyond mean-field, but because it introduces Laughlin correlations by avoiding pair states with the lowest allowed relative angular momentum $\mathcal{R} = 2\ell - L_2$. The allowed angular momentum multiplets which avoid pair states with $\mathcal{R} = 1$ form a subset of the set of multiplets $G_{N\ell}(L)$ that can be formed from N Fermions in a shell of angular momentum ℓ. This subset avoids the largest repulsion and has the lowest energy. Adiabatic addition of Chern-Simons flux introduces Laughlin correlations without the necessity of introducing an irrelevant mean-field energy scale $\hbar\omega_c = \nu^{-}\hbar\omega$.

Jain's sequence of filled CF shells does not require an interaction between CF quasiparticles. The incompressibility results from the energy required to create a QE-QH pair in the integrally filled CF state. Haldane's hierarchy of IQL states was based on the implicit assumption that the residual interaction between QPs was sufficiently similar to the Coulomb interaction between electrons in LL0 that the QPs would form their own Laughlin correlated daughter states.

The experiment of Pan et al. showed that neither Jain's CF picture nor Haldane's hierarchy was the whole story. Residual pair interactions between QPs had been determined up to an overall constant (unimportant for QP correlations). This pseudopotential $V_{\mathrm{QP}}(L_2)$ could be used to determine the spectrum of daughter states containing N_{QP} quasiparticles in a partially filled QP shell. Qualitatively correct results can be expected when $V_{\mathrm{QP}}(L_2)$ is small compared to the energy necessary to create a QE-QH pair in the IQL state. When the CF picture was reapplied to the QPs, the Haldane hierarchy of all odd denominator fractions resulted. Numerical calculations demonstrated that this CF hierarchy scheme of Laughlin correlated QPs at each level did not always work, because $V_{\mathrm{QP}}(L_2)$ was not sufficiently similar to $V_0(L_2)$, the pseudopotential for electrons in LL0.

Our study of when the MF CF picture is valid justifies Jain's ideas about integrally filled CF QP levels, but shows that higher generations of CF states resulting from CF QP interactions do not always support Laughlin correlations. Laughlin correlation among QPs is necessary to obtain a higher generation of CFs.

Acknowledgement
A number of graduate students, past doctoral research associates, and visiting faculty at the University of Tennessee made important contributions to the research reported in this paper. Among them are X. M. Chen, P. Sitko, A. Wojs, P. Hawrylak, K. S. Yi, Jennifer Quinn, and R. Wooten. Special thanks are due to K. S. Yi, who carefully read, corrected errors, and prepared this manuscript for submission.

References

[1] Jain J K 1989 *Phys. Rev. Lett.* **63** 199
[2] Chen X M and Quinn J J 1994 *Solid State Commun.* **92** 865
[3] Wojs A and Quinn J J 2000 *Phys. Rev.* B. **61** 2846
[4] Sitko P, Yi K S and Quinn J J 1997 *Phys. Rev.* B **56** 12417
[5] See, for example, Gasiorowicz S 1974 *Quantum Physics* (John-Wiley & Sons, New York)
[6] von Klitzing K, Dorda G and Pepper M 1980 *Phys. Rev. Lett.* **45** 494
[7] Haldane F D M 1983 *Phys. Rev. Lett.* **51** 605; Haldane F D M and Rezayi E H 1985 *Phys. Rev. Lett.* **54** 237
[8] See, for example, Quinn J J and Yi K S 2009 *Solid State Physics* (Springer-Verlag, Berlin Heidelberg)
[9] Tsui D C, Stormer H L and Gossard A C 1982 *Phys. Rev. Lett.* **48** 1559
[10] Laughlin R B 1983 *Phys. Rev. Lett.* **50** 1395
[11] Jain J K 1990 *Phys. Rev.* B **41** 7653
[12] Wojs A and Quinn J J 1999 *Solid State Commun.* **110** 45
[13] Benjamin A T, Quinn J J, Quinn J J and Wojs A 2001 *J. Combin. Theory* Series A **95** 390
[14] Quinn J J and Wojs A 2000 *Physica* E **6** 1

[15] Quinn J J and Wojs A 2000 *J. Phys.: Condens. Matt.* **12** R265
[16] Quinn J J, Wojs A, Yi K S and Simion G 2009 *Phys. Rep* **481** 29; We have used $2\ell^* = 2\ell - 2(N-1)$, $n_{\mathrm{QE}} = N - (2\ell^* + 1)$, and $\ell_{\mathrm{QE}} = \ell^* + 1$ in determining $(2\ell_{\mathrm{QE}}, n_{\mathrm{QE}})$.
[17] Pan W, Xia J-S, Shvarts V, Adams D E, Stormer H L, Tsui D C, Pfeifer L N, Baldwin K W and West K W 1999 *Phys. Rev. Lett.* **83** 3530
[18] Goerbig M, Lederer P and Smith C M 2006 *Physica* E **34** 1 and 2004 *Phys. Rev.* B **69** 155324; Lopez A and Fradkin E 2004 *Phys. Rev.* B **69** 155322; Smet J H 2003 Nature **422** 391
[19] Lee S Y, Scarola V W, and Jain J K 2002 *Phys. Rev.* B **66** 085336
[20] Wojs A, Wodzinski D and Quinn J J 2006 *Phys. Rev.* B **74** 035315
[21] Moore G and Read N 1991 *Nucl. Phys.* B **360** 362
[22] Greiter M, Wen X-G and Wilczek 1991 *Phys. Rev. Lett.* **66** 3205; 1992 *Nucl. Phys.* B **374** 507
[23] Samkharadze N, Arnold I, Pfeiffer L N, West K W and Csathy G A 2015 *Phys. Rev.* B **91** 081109

Journal of Physics: Conference Series **702** (2016) 012007 doi:10.1088/1742-6596/702/1/012007

Correlation diagrams: an intuitive approach to correlations in quantum Hall systems

S. B. Mulay, J. J. Quinn†, M. A. Shattuck

The University of Tennessee, Knoxville, Tennessee 37996, USA

E-mail: †jjquinn@utk.edu

Abstract. A trial wave function $\Psi(1, 2, \ldots, N)$ of an N electron system can always be written as the product of an antisymmetric Fermion factor $\mathcal{F}\{z_{ij}\} = \prod_{i<j} z_{ij}$, and a symmetric correlation factor $\mathcal{G}\{z_{ij}\}$. $\mathcal{F}$ results from the Pauli principle, and $\mathcal{G}$ is caused by Coulomb interactions. One can represent $\mathcal{G}$ diagrammatically [1] by distributing N points on the circumference of a circle, and drawing appropriate lines representing correlation factors (cfs) z_{ij} between pairs. Here, of course, $z_{ij} = z_i - z_j$, where z_i is the complex coordinate of the ith electron. Laughlin correlations for the $\nu = 1/3$ filled incompressible quantum liquid (IQL) state contain two cfs connecting each pair (i, j). For the Moore-Read state of the half-filled excited Landau level (LL), with $\nu = 2 + 1/2$, the even value of N for the half-filled LL is partitioned into two subsets A and B, each containing $N/2$ electrons [2]. For any one partition (A, B), the contribution to $\mathcal{G}$ is given by $\mathcal{G}_{AB} = \prod_{i<j\in A} z_{ij}^2 \prod_{k<\ell\in B} z_{k\ell}^2$. The full $\mathcal{G}$ is equal to the symmetric sum of contributions $\mathcal{G}_{AB}$ over all possible partitions of N into two subsets of equal size. For Jain states at filling factor $\nu = p/q < 1/2$, the value of the single particle angular momentum ℓ satisfies the equation $2\ell = \nu^{-1}N - C_\nu$, with $C_\nu = q + 1 - p$. The values of $(2\ell, N)$ define the function space of $\mathcal{G}\{z_{ij}\}$, which must satisfy a number of conditions. For example, the highest power of any z_i cannot exceed $2\ell + 1 - N$. In addition, the value of the total angular momentum L of the lowest correlated state must satisfy the equation $L = (N/2)(2\ell+1-N) - K_\mathcal{G}$, where $K_\mathcal{G}$ is the degree of the homogeneous polynomial generated by $\mathcal{G}$. Knowing the values of L for IQL states (and for states containing a few quasielectrons or a few quasiholes) from Jain's mean field CF picture allows one to determine $K_\mathcal{G}$. The dependence of the pair pseudopotential $V(L_2)$ on pair angular momentum L_2 suggests a small number of correlation diagrams for a given value of the total angular momentum L. Correlation diagrams and correlation functions for the Jain state at $\nu = 2/5$ and for the Moore-Read states will be presented as examples. The generalizations of the method of selecting $\mathcal{G}$ from small to larger systems will be discussed.

1. Introduction

In this paper, we study correlations resulting from Coulomb interactions in fractional quantum Hall systems. We introduce correlation diagrams to guide in the selection of the correlation functions caused by interactions. Electrons are represented by points located at positions z_i in the complex plane, and there are correlation lines connecting pairs of electrons. A correlation line connecting particles i and j represents a correlation factor (cf) $z_{ij} = z_i - z_j$. Although our correlation diagrams appear to resemble chemical bonds, they are just the opposite. A factor z_{ij}^m forbids the pair (i, j) from having a separation smaller than $m^{1/2}\lambda$, where $\lambda = (\hbar c/eB)^{1/2}$ is the magnetic length. An N electron system can be partitioned into subsets $(A, B, C, \ldots)$; one example is $(N) \to (N/2, N/2)$. There can be different numbers of cfs between pairs belonging to different subsets, and still different numbers between particles in different subsets. The number of cf lines associated with a particular partition can be determined. The subgroup of the full symmetric group associated with the conjugacy class of the partition is used to obtain the full symmetric correlation function. Our objective is to use correlation diagrams to gain new insights into correlations in strongly interacting many-body systems. New electronic correlation functions are obtained for states containing a few quasielectrons (QEs) in a partially filled QE shell, as well as for the incompressible quantum liquid states containing integrally filled QE shells.

For weakly interacting many-body systems, the interaction Hamiltonian H_I can be treated as a perturbation acting on energy eigenfunctions of a non-interacting Hamiltonian H_0. For strongly interacting systems, this standard many-body perturbation approach [3] is not applicable because the interaction energy is much larger than the single particle energy scale. The fractional quantum Hall (FQH) effect [4] is the ultimate example of strongly interacting many-body systems. In determining the energy spectrum of the interacting system at very large values of the applied magnetic field B_0, there is only one relevant energy scale, the Coulomb scale $V_c = e^2/\lambda$, where $\lambda = (\hbar c/eB_0)^{1/2}$ is the magnetic length. The non-interacting single particle states [5] for an electron confined to the x-y plane have eigenvalues $\varepsilon_{nm} = \hbar\omega_c[n+\frac{1}{2}(1+m+|m|)]$, where $\omega_c = eB_0/\mu c$ is the electron cyclotron frequency, $m = 0, \pm1, \pm2, \ldots$, and n is a non-negative integer. The lowest energy level (Landau level LL0) has $n = 0$, and m equal to a negative integer or zero. For a disk of finite area $\mathcal{A}$, the allowed values of m for the LL0 are $\{0, -1, -2, \ldots, -N_\phi\}$, where $N_\phi = \mathcal{A}B_0(\hbar c/e)^{-1}$ is the number of flux quanta of the applied magnetic field B_0 passing through the sample. Each of the $N_\phi + 1$ single particle states has the same energy $\frac{1}{2}\hbar\omega$. The non-interacting eigenfunction can be expressed in terms of a complex coordinate $z = x - iy$ of the electron as $\phi(z) \propto z^m$. Then, for LL0, $m \in g_0 \equiv \{0, +1, \ldots, +N_\phi\}$. Antisymmetrized products of N functions $\phi_m(z)$ selected from the set g_0 form the function space $(2\ell, N)$ of the LL0. Here we use 2ℓ in place of N_ϕ for convenience. Because the particles are Fermions, an N electron trial wave function can be written as a ubiquitous Gaussian weighting factor $e^{-\sum_k |z_k|^2/(4\lambda^2)}$, (which is often not explicitly written but is understood), multiplied by the product of an antisymmetric Fermion factor $\mathcal{F}\{z_{ij}\} = \prod_{i<j} z_{ij}$ caused by the Pauli exclusion principle, and a symmetric correlation function $\mathcal{G}\{z_{ij}\}$ caused by Coulomb interactions. Here, $z_{ij} = z_i - z_j$, and we often refer to it as a *correlation factor* (cf), even when it is caused by the Pauli principle and not by Coulomb correlations.

2. Electron correlations

Laughlin [6] realized that if the interacting electrons could avoid the most strongly repulsive pair states, an incompressible quantum liquid (IQL) state could result. He suggested a trial wave function for a filling factor ν (defined as $\frac{N}{2\ell+1}$) equal to the reciprocal of an odd integer n, in which the correlation function $\mathcal{G}_n(z_{ij})$ was given by $\prod_{i<j} z_{ij}^{n-1}$. This function is symmetric and avoids all pair states with relative pair angular momentum smaller than n (or all pair separations smaller than $r_n = n^{1/2}\lambda$). One can represent this Laughlin correlation function diagrammatically by distributing N dots, representing N electrons on the circumference of a circle, and drawing double lines, representing two correlation factors (cfs) connecting each pair. Thus, there are $2(N-1)$ cf factors in $\mathcal{G}\{z_{ij}\}$ emanating from each particle i. Adding $(N-1)$ cf factors emanating from each particle due to the Fermion factor $\mathcal{F}\{z_{ij}\}$ gives a total of $3(N-1)$ cfs emanating from each particle in the trial wave function Ψ. This number cannot exceed $N_\phi = 2\ell$ defining the function space $(2\ell, N)$ of the LL0.

The other well-known trial wave function is the Moore-Read [7] paired function describing the IQL state of a half filled spin polarized first excited Landau level (LL1). This wave function Ψ can be written in the form $\Psi = \mathcal{F} \cdot \mathcal{G}_{\rm MR}$, where the correlation function is taken as $\mathcal{G}_{\rm MR} = \mathcal{F}\{z_{ij}\}Pf(z_{ij}^{-1})$. The second factor is called the Pfaffian of z_{ij}^{-1}. It can be expressed as

[7, 8]

$$Pf(z_{ij}^{-1}) = \hat{\mathcal{A}} \prod_{i=1}^{N/2} (z_{2i-1} - z_{2i})^{-1},$$

where $\hat{\mathcal{A}}$ is an antisymmetrizing operator and the product is over pairs of electrons. There has been considerable interest in the Moore-Read paired state and its generalizations [9] based on rather formidable conformal field theory. We propose a simple intuitive picture of Moore-Read correlations with the hope that it might lead to new insight into correlations in strongly interacting many-body systems.

For the simple case of an $N = 4$ particle system, the Pfaffian can be expressed as

$$Pf(z_{ij}^{-1}) = \hat{\mathcal{A}}\{(z_{12}z_{34})^{-1}\} = \left[(z_{12}z_{34})^{-1} - (z_{13}z_{24})^{-1} + (z_{14}z_{24})^{-1}\right]. \quad (1)$$

The product of $\mathcal{F}\{z_{ij}\}$ and $Pf(z_{ij}^{-1})$ gives for the Moore-Read correlation function of

$$\mathcal{G}_{\text{MR}}\{z_{ij}\} = z_{13}z_{14}z_{23}z_{24} - z_{12}z_{14}z_{23}z_{34} + z_{12}z_{13}z_{24}z_{34}. \quad (2)$$

The correlation diagram for $\mathcal{G}_{\text{MR}}\{z_{ij}\}$ contains four points with a pair of cfs emanating from each particle i going to different particles j and k. There are three distinct diagrams shown in Fig. 1. Note that $\mathcal{G}_{\text{MR}}$ is symmetric under permutation, as it must be, since it is a product of two antisymmetric functions $\mathcal{F}\{z_{ij}\}$ and $Pf\{z_{ij}^{-1}\}$.

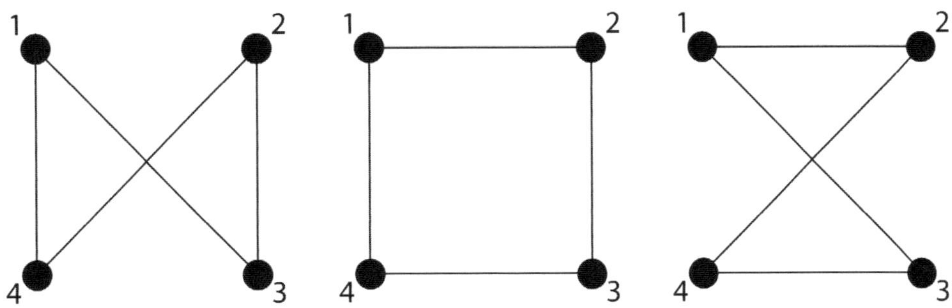

Figure 1. Moore-Read correlation diagram for $N = 4$. Dots represent particles, and solid lines represent cfs z_{ij}. $\mathcal{G}_{\text{MR}}$ is the symmetric sum of the three diagrams and is given by Eq. (2).

A simpler, but seemingly different, correlation is the quadratic function given by $\mathcal{G}_{\text{Q}} \equiv \hat{\mathcal{S}}(z_{12}^2 z_{34}^2)$, where $\hat{\mathcal{S}}$ is a symmetrizing operator. The correlation diagram for $\mathcal{G}_{\text{Q}}\{z_{ij}\}$ is shown in Fig. 2. $\mathcal{G}_{\text{MR}}$ and $\mathcal{G}_{\text{Q}}$ are clearly different. However, when they are expressed as homogeneous polynomials in the independent variables z_1 to z_4 (by simple multiplication), the two polynomials are the same up to normalization constant. The same was true for the an $N = 6$ particle system, leading to the conjecture that $\mathcal{G}_{\text{MR}}\{z_{ij}\}$ was equivalent to $\mathcal{G}_{\text{Q}}\{z_{ij}\}$ for all N. This conjecture was proved by our group before we discovered that Cappelli et al. [9] had already shown the equivalence.

There are several advantages to the use of $\mathcal{G}_{\text{Q}}$. First, it is simpler to partition N into two subsets of $N/2$, e.g., $\{1, 2, \ldots, N/2\} = A$ and $\{N/2 + 1, \ldots, N\} = B$, and define $g_{AB} = g_A g_B = \prod_{i<j\in A} z_{ij}^2 \prod_{k<\ell\in B} z_{k\ell}^2$. Then the full correlation function can be written as $\hat{\mathcal{S}}_N\{g_{AB}\}$, where $\hat{\mathcal{S}}_N$ symmetrizes g_{AB} over all N particles. This symmetrization is equivalent to summing g_{AB} over all possible partitions of N into two equal size subsets A and B. In Fig. 3, we show the contribution to $\mathcal{G}_{\text{Q}}$ for $N = 8$ particles for one partition in which $A = \{1, 3, 5, 7\}$ and $B = \{2, 4, 6, 8\}$.

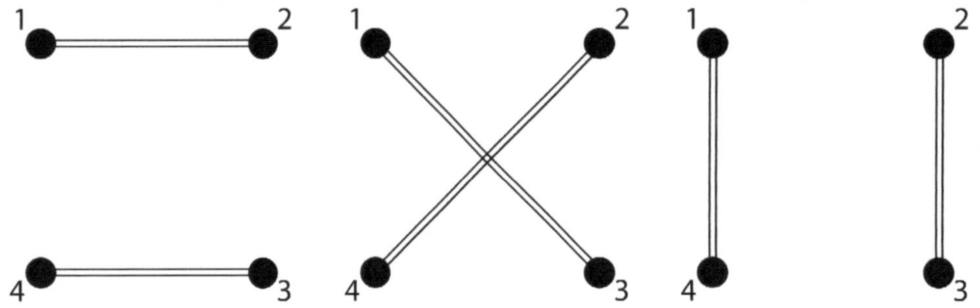

Figure 2. Quadratic correlation functions. A double line represents z_{ij}^2, the square of a cf. $\mathcal{G}_\mathrm{Q}$ is the sum of the contributions from the three diagrams.

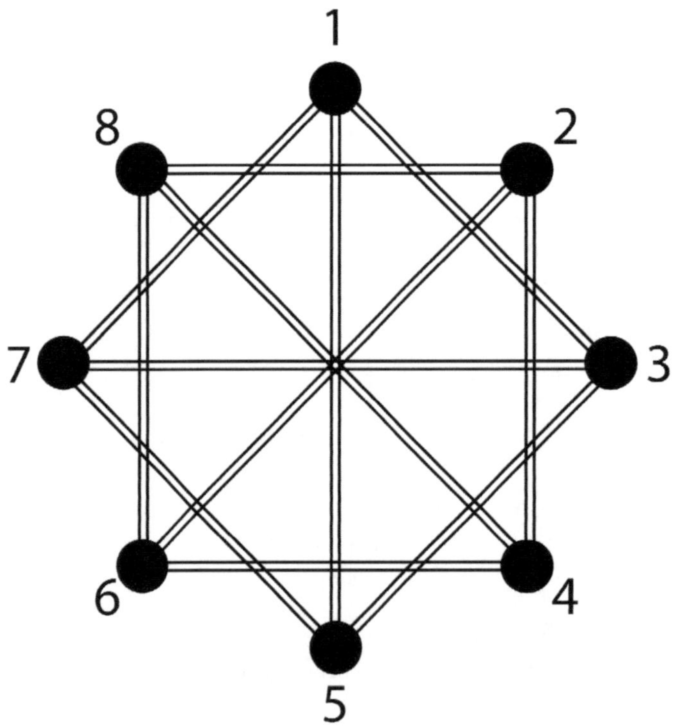

Figure 3. Correlation diagram for $\mathcal{G}_\mathrm{Q}\{z_{ij}\}$ in an eight electron system due to the partition $A = \{1, 3, 5, 7\}$ and $B = \{2, 4, 6, 8\}$. The full correlation function is the sum over all distinct partitions into subsets A and B, each containing $N/2 = 4$ particles. The trial wave function is $\Psi_Q(1, 2, \ldots, 8) = \mathcal{F}\{z_{ij}\}\mathcal{G}_\mathrm{Q}\{z_{ij}\}$.

3. Jain's composite Fermion approach

Jain [10] introduced a composite Fermion (CF) picture by attaching to each electron (via a gauge transformation) a flux tube which carried an even number $2p$ of magnetic flux quanta. This *Chern-Simons* (CS) flux has no effect on the classical equations of motion since the CS magnetic

Journal of Physics: Conference Series **702** (2016) 012007 doi:10.1088/1742-6596/702/1/012007

field $\mathbf{b}(\mathbf{r}) = 2p\phi_0 \sum_i \delta(\mathbf{r} - \mathbf{r}_i)\hat{z}$ vanishes at the position of each electron (it is assumed that no electron senses its own CS flux). Here $\phi_0 = \hbar c/e$ is the quantum of flux, and the sum is over all electron coordinates $\mathbf{r}_i$. The classical Lorentz force on the ith electron due to the CS magnetic field is $(-e/c)\mathbf{v}_i \times \mathbf{b}(\mathbf{r}_i)$ and $\mathbf{b}(\mathbf{r}_i)$ caused by the CS flux on every j not equal to i vanishes at the position $\mathbf{r}_i$. The CF model results in a much more complicated interaction Hamiltonian, but simplification results from making a mean field (MF) approximation in which the CS flux and the electron charge are uniformly distributed over the entire sample. The average electronic charge $-eN/\mathcal{A}$ is canceled by the fixed background of positive charge introduced to make the total charge vanish. This MF CF approximation results in a system of N non-interacting CFs (CF=electron plus attached flux tube) moving in an effective magnetic field $b^* = \nu b$. An effective CF filling factor ν^* was introduced satisfying the equation

$$\nu^{*-1} = \nu^{-1} - 2p. \tag{3}$$

This resulted in a filled CF level when ν^* was equal to an integer ($\nu^* = n = \pm 1, \pm 2, \ldots$) and a IQL daughter state at $\nu = n(1 + 2pn)^{-1}$. This Jain sequence of states was the most robust set of fractional quantum Hall states observed in experiments.

Making use of Haldane's spherical geometry [11, 12, 13], Chen and Quinn [14] introduced an *effective CF angular momentum* ℓ^* satisfying the relation $\ell_0^* = \ell - p(N-1)$, where $2p$ is the number of CS flux quanta per electron. The lowest CF Landau level (CF LL0) could hold $(2\ell^* + 1)$ CFs. There were $n_{\text{QE}} = N - (2\ell^* + 1)$ composite Fermion QEs of angular momentum $\ell_{\text{QE}} = \ell^* + 1$ or $n_{QH} = (2\ell^* + 1) - N$ CF QHs of angular momentum $\ell_{QH} = \ell^*$ if $2\ell^* + 1$ was not equal to N. This resulted in a lowest band of quasiparticle (QP) states separated by a gap from the higher energy quasi continuum. This allowed the total angular momentum states in this band to be determined by addition of angular momentum of n_{QP} quasiparticles each of angular momentum ℓ_{QP} using the rules for addition of Fermion angular momenta.

In Table 1, we summarize the results of Jain's MF CF picture of the low energy states of an $N = 4$ electron system for values of 2ℓ equal to 9, 8, 7, and 6. These correspond to the $\nu = 1/3$ filled IQL states and its excited states containing one, two, and three QEs. The table gives the values of ℓ, the single electron angular momentum, and the resulting values of $\ell_0^* = \ell - (N-1)$, the CF angular momentum; $n_{\text{QE}} = N - (2\ell_0^* + 1)$, the number of QEs; ℓ_{QE}, the QE angular momentum; $k_M = 2\ell - (N-1)$, the maximum number of correlation factor (cf) lines that can emanate from an electron in the correlation function $\mathcal{G}$; and the allowed values of the total angular momentum L which result.

ℓ	ℓ_0^*	n_{QE}	ℓ_{QE}	k_M	L
4.5	1.5	0	2.5	6	0
4	1	1	2	5	2
3.5	0.5	2	1.5	4	$0 \oplus 2$
3	0	3	1	3	0

Table 1. Values of ℓ for an $N = 4$ electron system and the values of ℓ_0^*, n_{QE}, ℓ_{QE}, k_M, and L which result.

It might seem surprising that Jain's very simple CF picture correctly predicts the angular momenta in the lowest band of states for any value of $(2\ell, N)$ which defines the function space of the many-body system. The initial guess that the Chern-Simons gauge interaction and the Coulomb interaction between fluctuations beyond the mean field canceled is certainly not correct. The gauge field interactions are proportional to $\hbar\omega_c$ which varies linearly with B, the applied magnetic field. However, the Coulomb interactions are proportional to e^2/λ (where λ is the magnetic length) and vary as $B^{1/2}$. The two energy scales cannot possibly cancel for all values of B. For very large values of B, only the Coulomb scale is relevant in determining the low energy band of states. Our group at the University of Tennessee demonstrated that the MF CF picture gave a valid description of the lowest band of states if the pair interaction energy $V(L_2)$ increased with increasing L_2 faster than the eigenvalue of $\hat{L}_2^2$, the square of the pair angular momentum [15].

Knowing this, and the occupancies of CF LLs from Jain's MF CF picture, makes it interesting to explore the correlations among the original electrons. We do this using correlation diagrams for small systems.

4. Correlation diagrams

We have already stated that Laughlin correlation can be described by drawing two cf lines between each pair $\langle i, j\rangle$. A cf line between i and j represents a correlation factor z_{ij}. The wave function $\Psi(1, 2, \ldots, N) = \mathcal{F}\{z_{ij}\}\mathcal{G}\{z_{ij}\}$ describing the IQL state at $\nu = 1/3$ will contain $3(N-1)$ cf lines emanating from each particle i. $(N-1)$ cf lines are associated with $\mathcal{F}\{z_{ij}\}$, leaving $2(N-1)$ cf lines associated with $\mathcal{G}\{z_{ij}\}$. The correlation diagram for a Laughlin $\nu = m^{-1}$ filling factor is simple, because every pair has exactly the same correlations. For other states, like a state with $n_{\rm QE}$ quasielectrons, the correlations are more complicated.

For simplicity, let's use as an example the $N = 4$ particle system with values of 2ℓ in the range $6 \leq 2\ell \leq 9$. The values of ℓ_0^*, $n_{\rm QE}$, $\ell_{\rm QE}$, k_M, and the total angular momentum L of the lowest energy bands for these states are given in Table 1. We define $K_{\mathcal{F}} = N(N-1)/2$ as the number of cf lines appearing in the Fermi function $\mathcal{F}\{z_{ij}\}$, and $K_{\mathcal{G}}$ as the number appearing in the correlation function $\mathcal{G}\{z_{ij}\}$. Knowing $N\ell$, $K_{\mathcal{F}}$, and the allowed values of total angular momentum L, we can determine $K_{\mathcal{G}}$ for each of the states listed in Table 1. For $\ell = 4.5$, 4, and 3 the corresponding values of $K_{\mathcal{G}}$ are 12, 8, and 6. For $\ell = 3.5$, there are two multiplets $L = 0$ ($K_{\mathcal{G}} = 8$) and $L = 2$ ($K_{\mathcal{G}} = 6$). We also know k_M from the table. With this information, we can construct correlation functions which have to be symmetric under permutation of a pair of particles. We show one correlation diagram for each of the values of 2ℓ. If it is not symmetric, we must apply $\hat{\mathcal{S}}_4$ on the function to symmetrize over all four particles.

For $(2\ell, N) = (9, 4)$ there is only a single diagram; it has 2 cfs connecting each pair of particles. For a one QE state, we must partition (4) into (3,1). The single particle i belongs to subset A and the other three j, k, ℓ belong to subset B. The latter subset has Laughlin correlations (z_{jk}^2) between each pair belonging to B. Particle i (in subset A) is the QE, and has single cf lines connecting it with two of the three particles in subset B. Fig. 4 shows one diagram. The diagram corresponds to $z_{12}z_{13}z_{23}^2z_{24}^2z_{34}^2$, and this function must be symmetrized by summing over all partitions of (4) into (3,1), i.e., including diagrams in which A can be 1, 2, 3, or 4. Notice that $k_M = 5$, $N\ell = 16$, and $K_{\mathcal{G}} = 8$, giving an $L = 2$ state for the single QE. For the two QE state with $(2\ell, N) = (7, 4)$, we partition (4) into (2,2). For example, let one partition be $A = (1, 2)$ and $B = (3, 4)$. One term in the correlation diagram is shown in Fig. 5. This diagram corresponds to $z_{12}^2z_{14}^2z_{23}^2$, and it must be symmetrized over all four particles. Notice that $k_M = 4$, $N\ell = 14$, and $K_{\mathcal{G}} = 6$, giving an $L = 2$. To obtain the $L = 0$ multiplet, we must add two more cfs. Fig. 6 shows one diagram for this case. It corresponds to a contribution $(z_{12}z_{23}z_{34}z_{41})^2$, and it must be symmetrized. Now $K_{\mathcal{G}} = 8$, and $L = 0$ results.

For $(2\ell, N) = (6, 4)$, we have three QEs with $k_M = 3$, and we can construct the diagram shown in Fig. 7. When symmetrized, it gives

$$\mathcal{G}\{z_{ij}\} = (z_{12}z_{34} + z_{13}z_{24})(z_{13}z_{42} + z_{14}z_{32})(z_{14}z_{23} + z_{23}z_{34}). \quad (4)$$

There is only one state of angular momentum $L = 0$, and the wave function $\Psi = \mathcal{F}\mathcal{G}$ obtained using equation (4) agrees exactly with that obtained by standard angular momentum addition.

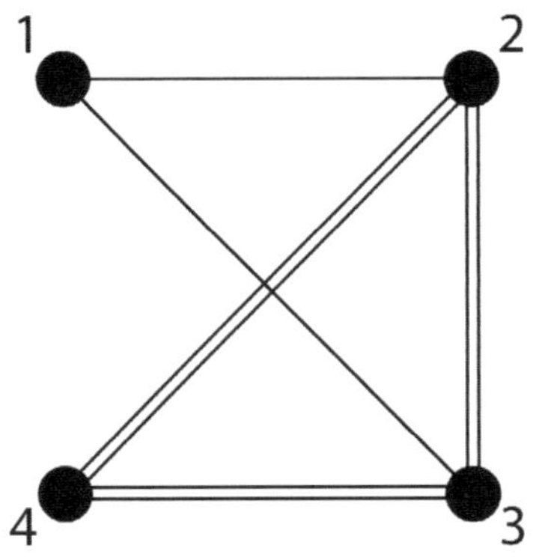

Figure 4. One contribution to $\mathcal{G}$ for $(2\ell, N) = (8, 4)$.

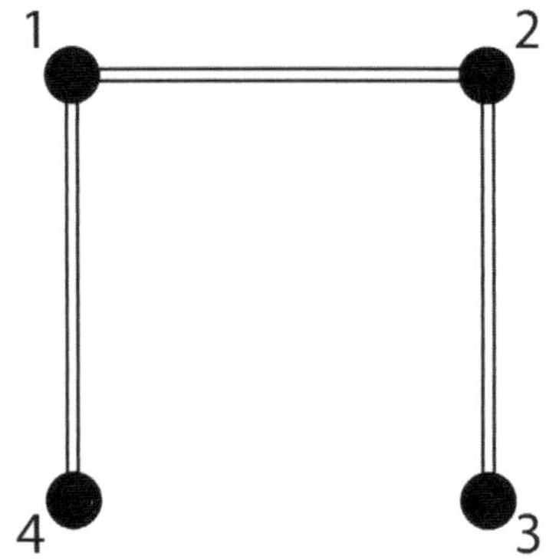

Figure 5. One contribution to $\mathcal{G}$ for $(2\ell, N) = (7, 4)$ that gives $L = 2$.

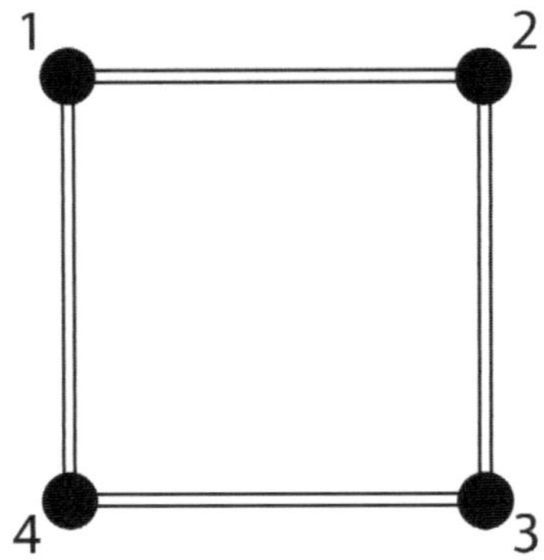

Figure 6. One contribution to $\mathcal{G}$ for $(2\ell, N) = (7, 4)$ that gives $L = 0$.

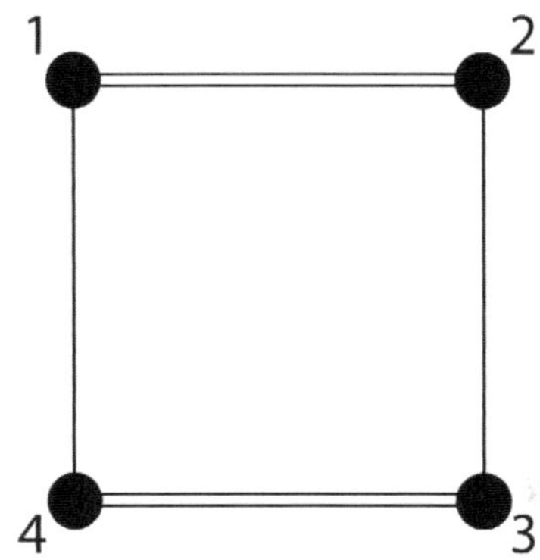

Figure 7. One contribution to $\mathcal{G}$ for $(2\ell, N) = (6, 4)$.

It is worth mentioning that there are three diagrams when Fig. 6 is symmetrized giving three terms $(z_{12}z_{23}z_{34}z_{41})^2$, $(z_{12}z_{13}z_{24}z_{34})^2$, and $(z_{13}z_{14}z_{23}z_{24})^2$. Their sum gives a symmetric $\mathcal{G}\{z_{ij}\}$. For Fig. 7, there are six diagrams giving $(z_{12}z_{34})^2 z_{13}z_{24}$, $(z_{12}z_{34})^2 z_{14}z_{23}$, $(z_{13}z_{24})^2 z_{12}z_{34}$, $(z_{13}z_{24})^2 z_{14}z_{23}$, $(z_{14}z_{23})^2 z_{12}z_{34}$, and $(z_{14}z_{23})^2 z_{13}z_{24}$. Plus or minus signs must be chosen for each term so that the resulting correlation function $\mathcal{G}\{z_{ij}\}$ is symmetric.

5. Thoughts on larger systems

Though we have used the $N = 4$ particle system as a simple example, it is not difficult to generalize to the case in which N is an arbitrary even integer. First, let us consider the Moore-Read state.

For the Moore-Read state, $2\ell = 2N - 3$, and we let $N = 2n$, where n is an integer. Then we have

(i) $2\ell = 4n - 3$

(ii) $k_M = 2n - 2$

(iii) Take a partition such as $A = (1, 2, \ldots, N/2)$, $B = (N/2 + 1, \ldots, N)$.
Take Laughlin correlations within subsets and write $g_A = \prod_{i<j\in A} z_{ij}^2$, $g_B = \prod_{k<\ell\in B} z_{k\ell}^2$, and

$$\mathcal{G} = \sum_{\text{all partitions}} g_A g_B. \tag{5}$$

Note that $N\ell = n(4n - 3)$, $K_{\mathcal{F}} = n(2n - 1)$, and $K_{\mathcal{G}} = n(2n - 2)$ giving $L = 0$. There are Laughlin correlations among the n particles in A and among the n particles in B, but no correlations between particles in different subsets. This is a simple intuitive way of fitting N particles into the function space $(2\ell, N)$ with maximum avoidance of the most repulsive pair states (ones with pair angular momentum $L_2 < 2\ell - 3$).

For the Jain state at $\nu = 2/5$, we can apply the same technique. In that case, we have

(i) $2\ell = 5n - 4$

(ii) $k_M = 3n - 3$

(iii) Partition into two subsets (A, B), as with the Moore-Read case.
Take g_A and g_B exactly as in that case.

(iv) Add a factor due to intersubset correlations to increase $K_{\mathcal{G}}$ so that an $L = 0$ state is produced despite the increase in the value of 2ℓ. For a partition such as $A = (1, 2, \ldots, n)$, $B = (n + 1, \ldots, 2n)$, define the intersubset correlation function

$$g_{AB} = \left(\prod_{i\in A}\prod_{j\in B} z_{ij}\right) \sum_{\sigma\in S_n} \prod_{i=1}^{n} z_{\sigma(i),n+i}^{-1}. \tag{6}$$

In (6), the first factor gives a product of n^2 correlation factors z_{ij}. The second factor is a sum of products of n factors of z_{ij}^{-1}. Define $G_{AB} = g_A g_B g_{AB}$ for a given partition (A, B). Then the full correlation function $\mathcal{G}$ in this case is the sum of the G_{AB} taken over all possible partitions.

6. Summary and conclusions

We have introduced a novel intuitive way of looking at electron correlations. For the very simple case of $N = 4$ particles, it gives, without numerical calculation, the exact ground state wave function of the Jain $\nu = 2/5$ filled state. For $2n = 6$ particles, the correlation function $\mathcal{G}\{z_{ij}\}$ is given by the sum over all partitions of $(2n)$ into two equivalent subsets (n, n) of G_{AB} (the contribution for the particular partition (A, B)). We have demonstrated that the trial wave function constructed with this correlation function has almost 99% overlap with the wave function [16] obtained by exact numerical diagonalization of the Coulomb interaction $H_I = \sum_{i<j} e^2/|z_{ij}|$. This result was presented as a poster at the International Conference on High Magnetic Fields in Semiconductors in 2014, and appeared in an unpublished workbook proceedings of the conference (available on the conference website).

Our method of obtaining the correlation functions for the Moore-Read and the Jain states is far more simple and intuitive than that of conformal field theory. It should be noted that the contributions of all partitions to the wave function are essentially the same since the wave function itself is antisymmetric. One need only count the number n_j of pairs which have j correlation factor lines and be sure that $\sum_j n_j = N(N - 1)/2$ and $\sum_j jn_j = K_{\mathcal{G}}$.

Acknowledgement
The authors thank Professors K. S. Yi and S. Johnston for assistance in preparing the manuscript, and acknowledge Dr. R. E. Wooten's numerical work cited in [16].

References
[1] Quinn J J, 2014 *Waves in Random and Complex Media* **24**, 279.
[2] In fact, it can also be partitioned into smaller subsets A, B, $C, \ldots$, down to $N/2$ subsets each containing a pair of electrons.
[3] See, for example, Abrikosov A A, Gorkov L P and Dzyaloshinski I E, 1963 *Methods of Quantum Field Theory in Statistical Physics* (Prestice-Hall Inc, Englewood Cliffs, NJ).
[4] Tsui D C, Stormer H L and Gossard A C, 1982 *Phys. Rev. Lett.* **48**, 1559.
[5] See, for example, Gasiorowicz S, 1974 *Quantum Physics* (John-Wiley & Sons, New York).
[6] Laughlin R B, 1983 *Phys. Rev. Lett.* **50**, 1395.
[7] Moore G and Read N, 1991 *Nucl. Phys.* B **360**, 362.
[8] Greiter M, Wen X-G and Wilczek, 1991 *Phys. Rev. Lett.* **66**, 3205; 1992 *Nucl. Phys.* B **374**, 507.
[9] Cappelli A, Georgiev L S and Todorov I T, 2000 *Proc. of Supersymmetries and Quantum Symmetries* (SQS'99, July 1999, Dubna), Ed. by Ivanov E, Krivonos S and Pashev A (JINR), p. 235.
[10] Jain J K, 1989 *Phys. Rev. Lett.* **63**, 199.
[11] Haldane F D M, 1983 *Phys. Rev. Lett.* **51**, 605; Haldane F D M and Rezayi E H, 1985 *Phys. Rev. Lett.* **54**, 237.
[12] There is a one-to-one correspondence between N electrons on a plane described by coordinates (r, ϕ) and N electrons on a sphere described by (ℓ, ℓ_z). For the plane, the z-component of angular momentum takes on the values $m = 0, 1, \ldots, N_\phi$ and the total z component of angular momentum is $M = \sum_{i=1}^{N} m_i$, where m_i is the z-component of angular momentum of a particle ($i = 1, 2, \ldots, N$). M is the sum of the *relative* angular momentum M_{R} and the center of mass angular momentum M_{CM}. On a sphere, the z-component of the single particle angular momentum is written as ℓ_z, and $|\ell_z| \leq \ell$, where ℓ is the angular momentum in the shell (or Landau level). The total angular momentum L is determined by addition of the angular momenta of N Fermions, each with angular momentum ℓ. N electron states are designated by $|L, L_z, \alpha\rangle$, where α is used to label different multiplets with the same value of L. It is apparent that $M = N\ell + L_z$, and one can show that $M_{\mathrm{R}} = N\ell - L$ and $M_{\mathrm{CM}} = L + L_z$. Therefore, for a state of angular momentum $L = 0$, M_{R} must be equal to $N\ell$. In general, the value of L for a given correlation function is given by the equation $L = N\ell - K_{\mathcal{F}} - K_{\mathcal{G}}$, where $K_{\mathcal{F}} = N(N-1)/2$ is the number of cf lines appearing in the Fermi function $\mathcal{F}$ and $K_{\mathcal{G}}$ is the number of cf lines in the correlation function $\mathcal{G}$.
[13] For $L = 0$, this relation was given by Fano G, Ortolani F and Colombo E, 1986 *Phys. Rev.* B **34**, 2670.
[14] Chen X M and Quinn J J, 1994 *Solid State Commun.* **92**, 865.
[15] Quinn J J, Wojs A, Yi K S and Simion G, 2009 *Phys. Rep.* **481**, 29 and references therein.
[16] Quinn J J, Wooten R E and Macek J H, 2014 *Workbook Proc. of the Int. Conf. on High Magnetic Fields in Semiconductors*, (unpublished, but available on conference website).

Many-body electron correlations in graphene

David Neilson[1], Andrea Perali[1] and Mohammad Zarenia[2]

[1] Dipartimenti di Fisica e di Farmacia, Università di Camerino, 62032 Camerino (MC), Italy
[2] Department of Physics, University of Antwerp, B-2020 Antwerpen, Belgium

E-mail: david.neilson@unicam.it

Abstract. The conduction electrons in graphene promise new opportunities to access the region of strong many-body electron-electron correlations. Extremely high quality, atomically flat two-dimensional electron sheets and quasi-one-dimensional electron nanoribbons with tuneable band gaps that can be switched on by gates, should exhibit new many-body phenomena that have long been predicted for the regions of phase space where the average Coulomb repulsions between electrons dominate over their Fermi energies. In electron nanoribbons a few nanometres wide etched in monolayers of graphene, the quantum size effects and the van Hove singularities in their density of states further act to enhance electron correlations. For graphene multilayers or nanoribbons in a double unit electron-hole geometry, it is possible for the many-body electron-hole correlations to be made strong enough to stabilise high-temperature electron-hole superfluidity.

1. Introduction

The Coulomb repulsion between conduction electrons in conventional metals and semiconductors affects their properties only through the relatively weak effects of linear screening and corrections to the values of the Landau Fermi liquid parameters. This is associated with the high densities of conduction electrons found in these materials. At such high densities, the Fermi energies are large or comparable to the average electron-electron interaction energies, and so the electrons behave as a weakly-interacting system. In such systems, many-body electron correlations are so weak that they play only a marginal role.

1.1. New phenomena for strongly correlated conduction electrons

If the density of conduction electrons could be lowered sufficiently to make electron interactions dominate over Fermi energies, a wealth of interesting new quantum phenomena driven by strong many-body electron correlations are predicted to appear. These phenomena include:

- Wigner crystal of electrons
- Charge density waves and other striped ground states
- Metal-insulator transition
- Quantum glass in the presence of defects
- Coherent superconductor and electron-hole superfluid quantum states
- Crossover from BCS superconductivity to Bose-Einstein condensation

1.2. When are many-body correlations expected to be important ?

A two-dimensional (2D) electron layer is expected to have correlations that are stronger than the correlations in a corresponding three- dimensional system at the same density, because of the smaller kinetic energy contributions. The dimensionless parameter r_s provides a measure of when electron correlations will be important. It is defined as $r_s = \langle PE\rangle/\langle KE\rangle$, and represents the relative importance of the average kinetic energy at zero temperature, $\langle KE\rangle = \langle \hbar^2\nabla^2/2m^*\rangle \simeq \langle \hbar^2/(m^* r_0^2)\rangle$, to the average strength of the Coulomb potential energy from electron repulsion, $\langle PE\rangle = \langle e^2/\kappa r_0\rangle$, where m^* is the electron effective mass and κ the dielectric constant of the background medium. For electrons of density n in a 2D layer, the average spacing between the electrons $r_0 = 1/\sqrt{\pi n}$. In a 2D system with parabolic dispersion of the energy bands, $r_s = r_0/a_B^*$, where the effective Bohr radius for the system is $a_B^* = \kappa\hbar^2/(e^2 m^*)$. Thus at sufficiently large r_0 (low electron densities), $r_s \gg 1$, and the electron Coulomb interactions will dominate over their Fermi energy.

2. Strongly correlated conduction electron systems

2.1. 2D Electron Liquid in Si MOSFETs and GaAs heterostructures

In 2D systems, the new phenomena driven by strong correlations are predicted to appear at very low densities, generally only for $r_s \gtrsim 10$ [1]. For many years quasi-2D electron layers at the interfaces of Si MOSFET devices and in narrow quantum wells in GaAs heterostructures (Fig. 1) have shown great promise as systems to access this low density region [2]. However even in the highest quality semiconductors with extremely high mobilities, residual defects and impurities still block much access to this interesting region, since when the electron density is lowered by metal gates or by doping, the samples always eventually become insulating. This electron freeze-out occurs when the remaining conduction electrons can no longer screen residual charged impurities and they become trapped by the impurities. The finite widths of the electron layers or quantum wells is a further challenge, as quantum wells must be wider than 10-15 nm in order to retain their good conduction properties. Finite widths mean weaker Coulomb interactions between electrons.

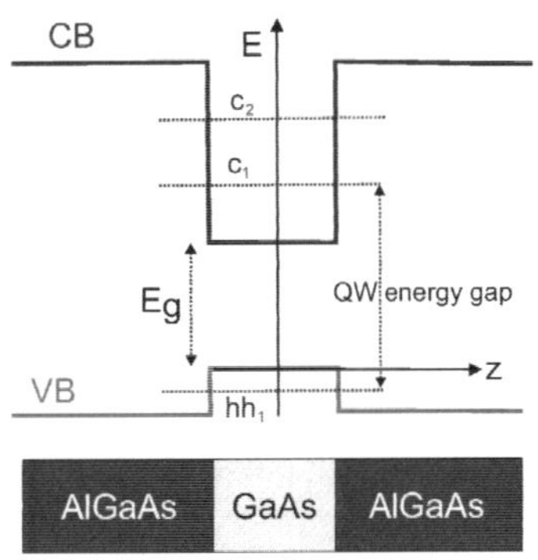

Figure 1. Quantum well in a GaAs heterostructure.

2.2. Graphene

In contrast to the quasi-2D electron layers in semiconductor systems, a graphene sheet is atomically flat and hence strictly 2D. There are no finite-width effects to weaken the Coulomb interactions. Levels of defects in graphene are extremely low so that electron freeze-out should be postponed to much lower electron densities.

2.2.1. Monolayer graphene

However nature has not been kind. Because the dispersion of the energy bands in monolayer graphene is linear at low energies, $E_{\pm}(k) = \pm\hbar v_F k$ (Fig. 2), the Fermi energy $E_F = \hbar v_F k_F = \hbar v_F \sqrt{\pi n}$ and $\langle KE \rangle$ depend only linearly on r_0^{-1}. A result of this is that the $r_s = e^2/(\kappa\hbar v_F)$ in monolayer graphene does not depend on the electron density. The Fermi velocity $v_F \simeq 10^6$ ms^{-1} and the dielectric constant of the substrate is typically $\kappa \sim 3$ [3], making the fixed value of r_s small, $r_s \sim 0.5$-0.7. Such a small value for r_s implies that the correlations of the conduction electrons in monolayer graphene will always remain weak [4].

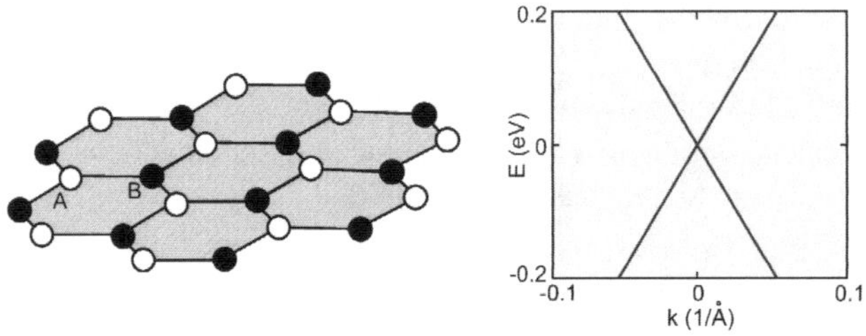

Figure 2. Monolayer graphene: atomic structure and linear energy bands at low energies.

2.2.2. Bilayer graphene

It should, nevertheless, be possible to access the strongly correlated region in graphene. One work-around is to substitute a graphene bilayer sheet in place of the graphene monolayer. A symmetrically biased graphene bilayer with AB stacking is a semiconductor with parabolic dispersion of the energy bands [5] (Fig. 3). With parabolic dispersion, the interaction parameter r_s once again increases with decreasing electron density as it does in a semiconductor system [6]. Furthermore, in contrast to monolayer graphene, if a perpendicular electric field is applied across bilayer graphene using a metal gate, the field will generate a tuneable energy gap between the conduction and valence bands.

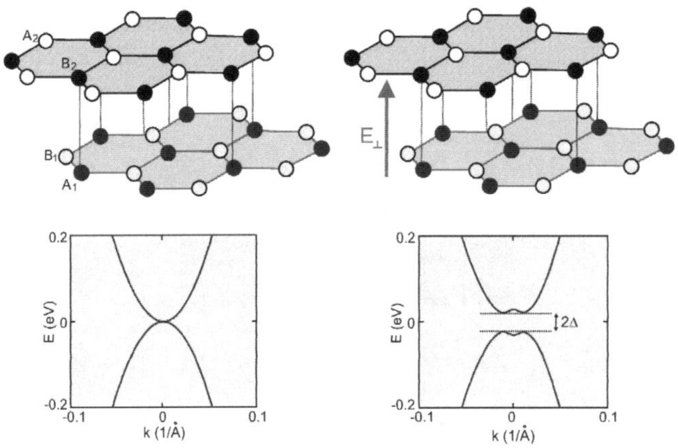

Figure 3. Bilayer graphene: atomic structure and parabolic energy bands at low energies. With an external perpendicular electric field $E_\perp$ there is an energy gap 2Δ between the valence and conduction bands.

The low density regime in bilayer graphene is dominated by disorder in current samples, making a lower limit for the electron density of $n \sim 10^{10}$ cm^{-2}, corresponding to a large maximum value for

$r_s = 23$. At very low densities a trigonal warping of the bands could transform the parabolic bands into sets of Dirac-like linear bands [7], but residual disorder will mask this effect, and it could be further reduced if necessary by applying an electric field to open up an energy band gap.

In summary, extreme high quality, atomically flat bilayer graphene sheets have tuneable electron densities and band gaps that should permit the bilayers to readily access new quantum phenomena predicted for strong electron correlations.

2.2.3. Few-layer graphene

Increasing the number of graphene layers in the sheet beyond bilayers greatly enhances the density of states (DOS), and this projects the sheets even more dramatically into the region of strong correlations at accessible densities [8]. Electron graphene multilayers should be able to access regions of phase space with very strong electron-electron interactions. For the lowest energy band in ABC stacked N-layer graphene, the energy dispersion of the conduction band is given by [9, 10],

$$E^{(N)}(k) = \left\{(\hbar v_F)^N / t^{N-1}\right\} k^N \,, \tag{1}$$

where $t \approx 400$ meV is the interlayer hopping term in few-layer graphene. Figure 4(a) shows $E^{(N)}(k)$ for $N = 1$ to 4. Using Eq. (1) to determine the Fermi energy E_F at density n, we obtain for N-layer graphene,

$$r_s = \langle PE\rangle / \langle KE\rangle = \left\{ \frac{e^2 t^{N-1}}{\kappa(\hbar v_F)^N \sqrt{\pi}^{N-1}} \right\} \frac{1}{n^{(N-1)/2}} \,. \tag{2}$$

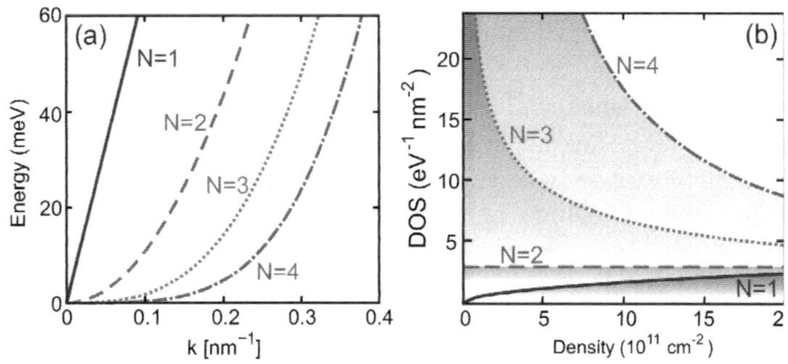

Figure 4. (a) Lowest positive energy band in monolayer ($N = 1$), bilayer ($N = 2$), trilayer ($N = 3$), and quadlayer ($N = 4$) graphene. (b) Comparison for $N = 1$ to $N = 4$ of the density of states at the Fermi energy $DOS^{(N)}(E_F)$ as function of electron density [8]. The van Hove power law singularities as $E_F \to 0$ lead to large enhancements of the DOS for trilayer and quadlayer graphene.

A consequence of the different energy dispersions $E^{(N)}(k)$ is that the dependence of the density of states on energy changes dramatically with the number of layers N,

$$DOS^{(N)}(E) = \frac{d\Omega^{(N)}}{dE} = \frac{2\pi}{N} \frac{t^{2(N-1)/N}}{(\hbar v_F)^2} E^{(2/N)-1} \,, \tag{3}$$

where $\Omega^{(N)}(k)$ is the volume in k-space of the N-layer sheet. Figure 4(b) shows the dependence of $DOS^{(N)}(E_F)$ at the Fermi energy on electron density n. For monolayer graphene $DOS^{(1)}(E_F)$ depends linearly on n, for bilayer graphene $DOS^{(2)}(E_F)$ is a constant, and for trilayer and quadlayer graphene $DOS^{(N)}(E_F)$ decreases with increasing n. Because of the van Hove singularities as $E \to 0$, the $DOS^{(3)}(E_F)$ and $DOS^{(4)}(E_F)$ for small densities are much larger than $DOS^{(1)}(E_F)$

Density (cm^{-2})	monolayer	bilayer	trilayer	quadlayer
5×10^{12}	0.7	1	2	3
1×10^{12}	0.7	3	8	29
5×10^{11}	0.7	4	17	83
1×10^{11}	0.7	8	86	930

Table 1. Values of the interaction parameter r_s for few-layer graphene as a function of electron density. Large values of $r_s \gg 1$ can lead to new phases driven by the strong electron correlations.

and $DOS^{(2)}(E_F)$. Eventually, at very high densities lying well outside our range of interest, the $DOS^{(3)}(E_F)$ and $DOS^{(4)}(E_F)$ are smaller than $DOS^{(1)}(E_F)$ and $DOS^{(2)}(E_F)$.

Table I compares the values of r_s for the typical electron densities found in graphene sheets for N-layer graphene, for $N = 1$ (monolayer) to $N = 4$ (quadlayer). The table shows that few-layer graphene offers dramatic opportunities for producing extremely strongly interacting electron systems at experimentally accessible densities.

Experimental realisation of few layer graphene is within the grasp of current technology since few-layer graphene sheets can be fabricated in large areas by both mechanical exfoliation [11, 12] and by chemical techniques [13, 14, 15] from graphite with controlled stacking order. References [16, 17, 18] are examples of experimental studies on electronic and transport properties in trilayer graphene.

2.2.4. *Graphene nanoribbons*

Electrons can be confined in nanoribbons that are only a few nanometres in width, etched on monolayers of graphene. The nanoribbons can have multiple energy subbands that are occupied. Their electronic properties depend on the type of edge termination [19]. We discuss here only armchair-edge terminated graphene nanoribbons (Fig. 5), since, unlike for zig-zag-edge terminated nanoribbons, the multiple subbands of armchair nanoribbons are parabolic around their minima and there is a semiconductor-like energy gap between conduction and valence bands (Fig. 5). Also, there is no valley degeneracy. These properties all act to diminish the effect of Coulomb screening. Uniform armchair graphene nanoribbons of widths much less than 10 nm have already been fabricated [20].

Electron correlations in armchair-edge terminated nanoribbons will be further boosted by the quantum confinement of the electrons along the nanoribbons, and also by quantum size effects and van Hove singularities in the quasi-one-dimensional density of states that are accessed if the Fermi energy is increased so it enters the bottom of a new subband.

Figure 5 shows the single-particle energy subbands j in the continuum model, $\epsilon_j(k_y) = (\sqrt{3}ta_0/2)\sqrt{k_y^2 + k_j^2}$, $j = 1, 2, \ldots$ for an armchair graphene nanoribbon of width $W = 2$ nm, where $t = 2.7$ eV is the intralayer hopping energy [21] and $a_0 = 0.24$ nm is the graphene lattice constant. We take the y-direction parallel to the nanoribbon, with the electrons confined in the transverse x-direction. The quantised wave-number for the j-subband in the x-direction is $k_j = \lfloor j\pi/W \rfloor - \lfloor 4\pi/(3\sqrt{3}a_0) \rfloor$. The lower panels in Figure 5 show the corresponding density of states DOS(E). The van Hove singularities are clearly visible at the bottom of each subband.

2.3. *Graphene in a periodic magnetic field*

A periodic magnetic field applied perpendicular to the graphene monolayers can preserve the isotropic Dirac cones of the monolayer energy bands while reducing the slope of the Dirac cones [23, 24, 25, 26]. We represent the magnetic field perpendicular to the monolayer as a one-dimensional array with period $2d$ of successive rectangular magnetic barriers and wells of height $B_z = \pm B$ and dimensionless width $d = d_B/\ell_B$, where $\ell_B = \sqrt{\hbar c/eB}$ is the magnetic length. Since the average of the magnetic flux across a unit cell of the periodic field is fixed at zero, the main effect of the magnetic field is to modify the

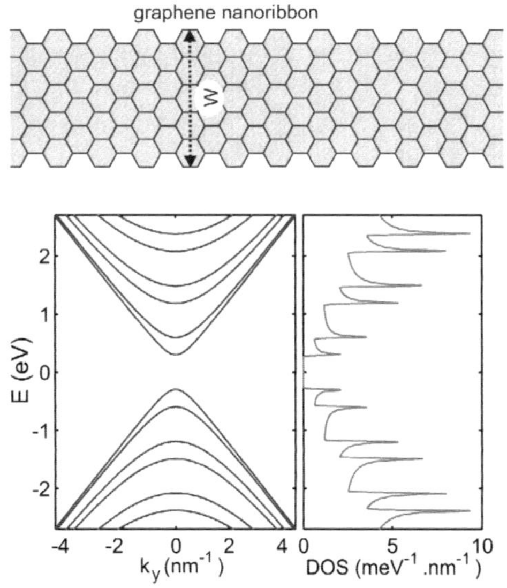

Figure 5. Top: armchair-edge terminated graphene nanoribbon of width W. Lower panels: lowest single-particle energy subbands $\epsilon_j(k_y), j = 1, 2, \ldots$ for an armchair nanoribbon of width $W = 2$ nm, and corresponding density of states DOS(E) in the nanoribbon [22]. Van Hove singularities are visible at the bottom of each subband.

electronic band structure of the graphene monolayers. In this field, the single-particle energy dispersion of the monolayer graphene remains linear, but the velocity $\alpha_d v_F$ is less than the original Fermi velocity v_F, so

$$\epsilon(\mathbf{k}) = \pm\hbar(\alpha_d v_F)|\mathbf{k}|\,(1 + \delta(\mathbf{k}))\ . \tag{4}$$

The non-linear correction term $\delta(\mathbf{k})$ is small, with $|\delta(\mathbf{k})| \lesssim d_B^2 k_x^2/6$. The constant $\alpha_d \leq 1$, representing the decrease in the Fermi velocity, depends on d [23, 26],

$$\begin{aligned} \alpha_d &\simeq 1 - d^4/60 \qquad && d \ll 1 \\ \alpha_d &\simeq \tfrac{2d}{\sqrt{\pi}} e^{-d^2/4} \qquad && d \gg 1\ . \end{aligned} \tag{5}$$

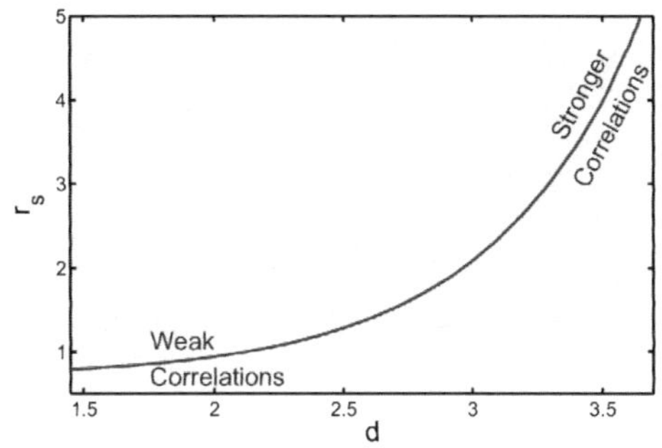

Figure 6. Variation of the interaction parameter r_s for monolayer graphene in a perpendicular magnetic field of periodicity $2d = 2d_B/\ell_B$ (see text). Without the magnetic field, r_s is a constant less than unity, and is independent of electron density.

Recalling for graphene monolayers that the density-independent interaction strength parameter is $r_s = e^2/(\kappa\hbar v_F)$, we see that an effect of the reduction of the Fermi velocity in Eq. 4 is to increase the value of the r_s parameter by a factor α_d^{-1}. Figure 6 shows that we can significantly increase r_s by using the periodic magnetic field to tune α_d with d.

3. Electron-hole superfluidity in graphene

Two graphene monolayers of electrons and holes separated by a very thin insulating barrier has been proposed to observe an electron-hole superfluid [27, 28, 29]. A hexagonal boron nitride (hBN) separation barrier as thin as 1 nm can efficiently insulate the two monolayers from each other [30]. However, theory suggests that because of strong screening of the electron-hole pairing interaction, an electron-hole superfluid can only occur when $r_s > 2.3$ [31]. Experiments confirm that the superfluid is not seen in two graphene monolayers [30], where we recall r_s is fixed independent of density at $r_s \sim 0.5$-0.7.

This poses a challenge as to whether new structures can be designed and fabricated using atomically thin crystals, structures in which a superfluid transition can be observed. The graphene sheets discussed in Sections 2.2.2 to 2.2.4 can all access the region $r_s \gg 2.3$, and hence they constitute promising candidates to generate this elusive electron-hole superfluid [6, 8, 32, 33].

In superfluid systems in graphene, it would be straightforward to access the BCS-BEC crossover and BEC regimes using the electric potential on metal gates and by tuning the sample parameters. This possibility opens up interesting new connections with the physics of ultracold fermions and high-T_c superconductors.

We have predicted the existence of electron-hole superfluidity in the graphene structures described in Sections 2.2.2 to 2.2.4. We find mean field zero temperature superfluid gaps that are consistently large, of the order of several hundred Kelvin. Unlike in three-dimensions, however, the superfluid phase transition temperature T_c in a two-dimensional system is not related linearly to the magnitude of its zero temperature superfluid gap. For quantum condensates in two-dimensions, an upper bound on the transition temperature is the Kosterlitz-Thouless temperature [34], $T_{KT} = (\pi/2)J(T_{KT})$, where $J(T)$ is the average kinetic energy of the Cooper pairs. At zero temperature, $J(0)$ is proportional to the superfluid density, $J(0) = \rho_s(0)/2$, and, since in mean field $\rho_s(T)$ depends very weakly on temperature for T small compared with the zero-temperature superfluid gap, for this case we may approximate $J(T)$ by $J(0)$ [35].

We find the resulting T_{KT} is typically of order 15 K for the electron-hole coupled graphene bilayers and the electron-hole coupled graphene monolayers in a periodic magnetic field, for sheet separations ~ 2 nm and typical experimental carrier densities, $n \sim 10^{12}$ cm^{-2}. Switching from coupled electron-hole bilayers to the coupled electron-hole quadlayers for the same sheet separation, typically doubles the T_{KT}. These results indicate that electron-hole superfluidity should be readily detectable in these graphene structures using existing technology.

4. Conclusions

With graphene and related atomically thin crystals, we are at an exciting many-body threshold to realise and exploit novel quantum phases with tuneable properties. High quality, atomically flat two-dimensional electron graphene sheets and quasi-one-dimensional electron graphene nanoribbons with tuneable electron densities and band gaps, should exhibit novel phenomena driven by strong many-body correlations. These phenomena are predicted when Coulomb repulsions between electrons are dominant over their Fermi energies, and include the Wigner crystal and charge density waves. In addition, certain configurations of double graphene sheets or nanoribbons, one doped with electrons and one with holes, should generate an electron-hole superfluid state characterised by a very large superfluid energy gap and relatively high transition temperature.

Acknowledgments
We thank Lucian Covaci and François Peeters (University of Antwerp), Alexander Hamilton (University of New South Wales), and Luca Dell'Anna (University of Padua) for useful discussions. We acknowledge support by the University of Camerino FAR project CESEMN (DN and AP), the Flemish Science Foundation (FWO-Vl) (MZ) and the University of Antwerp Research Fund (BOF) (MZ). The authors thank colleagues involved in the MultiSuper International Network (http://www.multisuper.org) for exchange of ideas connected with this work.

References

[1] Świerkowski L, Neilson D, and Szymański J 1991 Phys. Rev. Lett. **67**, 240; Rapisarda Francesco and Senatore Gaetano 1996 Aust. J. Phys. **49**, 161
[2] Zheng B, Croxall A F, Waldie J, Das Gupta K, Sfigakis F, Farrer I, Beere H E, Ritchie D A 2015 ArXiv:1511.08701
[3] Young A F, Dean C R, Meric I, Sorgenfrei S, Ren H, Watanabe K, Taniguchi T, Hone J, Shepard K L and Kim P 2012 Phys. Rev. B **85**, 235458
[4] Dahal Hari P, Joglekar Yogesh N, Bedell Kevin S, Balatsky Alexander V 2006 Phys. Rev. B **74**, 233405; Das Sarma S, Adam Shaffique, Hwang E H and Rossi Enrico 2011 Rev. Mod. Phys. **83**, 407
[5] McCann E and Fal'ko V I 2006 Phys. Rev. Lett. **96**, 086805
[6] Perali A, Neilson D and Hamilton A R 2013 Phys. Rev. Lett. **110**, 146803
[7] Castro Neto A H, Guinea F, Peres N M R, Novoselov K S and Geim A K 2009 Rev. Mod. Phys. **81**, 109
[8] Zarenia M, Perali A, Neilson D and Peeters F M 2014 Sci. Reports **4**, 7319
[9] Min H and MacDonald A H 2008 Phys. Rev. B **77**, 155416
[10] Katsnelson, M I 2012 *Graphene: Carbon in two dimensions* (New York: Cambridge University Press)
[11] Ferrari A C, Meyer J C, Scardaci V, Casiraghi C, Lazzeri M, Mauri F, Piscanec S, Jiang D, Novoselov K S, Roth S and Geim A K 2006 Phys. Rev. Lett. **97**, 187401
[12] Zhang Y, Small J P, Pontius W V and Kim Philip 2005 Appl. Phys. Lett. **86**, 073104
[13] Berger C, Song Z, Li T, Li X, Ogbazghi A Y, Feng R, Dai Z, Marchenkov A N, Conrad E H, First P N and de Heer W A 2004 J. Phys. Chem. B **108**, 19912
[14] Shih Chih-Jen, Vijayaraghavan A, Krishnan R, Sharma R, Han Jae-Hee, Ham Moon-Ho, Jin Zh, Lin Sh, Paulus G L C, Reuel N F, Wang Q H, Blankschtein D and Strano M S 2011 Nature Nanotechnology **6**, 439
[15] Mahanandia P, Simon F, Heinrich G and Nanda K K 2014 Chem. Commun. **50**, 4613
[16] Craciun M F, Russo S, Yamamoto M, Oostinga J B, Morpurgo A F and Tarucha S 2009 Nature Nanotech. **4**, 383
[17] Bao W, Jing L, Velas J, Aykol M, Cronin S B, Smirnov D, Koshino M, McCann E, Bockrath M and Lau C N 2011 Nature Phys. **7**, 948
[18] Mak Kin Fai, Shan Jie and Heinz Tony F 2010 Phys. Rev. Lett. **104**, 176404
[19] Brey L and Fertig H A 2007 Phys. Rev. B **75**, 125434
[20] Tapasztó L, Dobrik G, Lambin P and Biró L P 2008 Nature Nanotechnology **3**, 397
[21] Brey L and Fertig H A 2006 Phys. Rev. B **73**, 235411
[22] Zarenia M, Perali A, Peeters F M and Neilson D 2015 ArXiv:1601.06942
[23] Dell'Anna L and De Martino A 2009 Phys. Rev. B **79**, 045420; *ibid.* 2009 **80**, 089901(E)
[24] Snyman I 2009 Phys. Rev. B **80**, 054303
[25] Tan L Z, Park C -H and Louie S G 2010 Phys. Rev. B **81**, 195426
[26] Dell'Anna L and De Martino A 2011 Phys. Rev. B **83**, 155449
[27] Zhang C -H and Joglekar Y N 2008 Phys. Rev. B **77**, 233405; Lozovik Yu E and Sokolik A A 2008 JETP Lett. **87**, 55; 2010 Eur. Phys. J. B **73**, 195; Mink M P, Stoof H T C, Duine R A and MacDonald A H 2011 Phys. Rev. B **84**, 155409
[28] Min Hongki, Bistritzer Rafi, Su Jung-Jung and MacDonald A H 2008 Phys. Rev. B **78**, 121401(R)
[29] Bistritzer R, Min H, Su J -J and MacDonald A H 2008 ArXiv:cond-mat/0810.0331v1
[30] Gorbachev R V, Geim A K, Katsnelson M I, Novoselov K S, Tudorovskiy T, Grigorieva I V, MacDonald A H, Watanabe K, Taniguchi T and Ponomarenko L A 2012 Nat. Phys. **8**, 896
[31] Lozovik Yu E, Ogarkov S L and Sokolik A A 2012 Phys. Rev. B **86**, 045429
[32] Neilson D, Perali A and Hamilton A R 2014 Phys. Rev. B **89**, 060502(R)
[33] Dell'Anna L, Perali A, L Covaci L and Neilson D 2015 Phys. Rev. B **92** 220502(R)
[34] Kosterlitz J M and Thouless D J 1973 J. Phys. C: Sol. State Phys., **6**, 1181
[35] Benfatto L, Capone M, Caprara S, Castellani C, and Di Castro C 2008 Phys. Rev. B **78**, 140502(R); Benfatto L and Sharapov S G 2006 Low Temp. Physics **32**, 533

On dissipative effects in volume plasmon response

P.-G. Reinhard[a], **E. Krotscheck**[b],**E. Suraud**[b,c,d]

[a]Institut für Theoretische Physik, Universität Erlangen, D-91058 Erlangen, Germany
[b]Physics Department, University at Buffalo, The State University New York, Buffalo, NY 14260, USA and
Institute for Theoretical Physics, JKU Linz, Linz, Austria
[c]Université de Toulouse, UPS; Laboratoire de Physique Théorique (IRSAMC), F-31062 Toulouse Cedex, France
[d]CNRS; LPT (IRSAMC), Université de Toulouse, F-31062 Toulouse Cedex, France

E-mail: `suraud@irsamc.ups-tlse.fr`

Abstract. We study the imortance of dissipation on the volume plasmon response in moderate size Sodium clusters. Volume plasmon response has been tentatively identified experimentally in Sodium clusters. An open question is whether the standard mean-field approach, based on Density Functional Theory (DFT), is sufficient to provide a proper description of such modes in a possibly non-linear regime of excitations. A conceptually simple approximation, based on a recently proposed quantum Relaxation Time Ansatz (RTA), is used to evaluate these effects. It turns out that the response is, in the explored cases, clearly dominated by the mean field, which allows the use of simplified theories that give access to larger systems.

1. Introduction

Finite fermions systems have for a long time been investigated by studying collective excitations [1]. In metal clusters, the optical response provides a key to understanding dynamical scenarios in a wide range of excitation energies [2, 3]. It has been studied especially in the low-energy/long-wavelength domain where the Mie surface plasmon plays a dominant role [4, 5]. Still, traces of volume plasmon excitations (associated with local density oscillations) have also been tentatively observed experimentally [6], along the lines of earlier theoretical investigations [7]. These results in metals complement similar investigations on volume plasmon modes in C_{60} [8, 9].

A standard way to analyze collective modes is to study the dynamic structure function $S(\mathbf{k}, \omega)$ of the system. Many investigations of $S(\mathbf{k}, \omega)$ have been carried out in the long-wavelength limit ($k \longrightarrow 0$), providing a good description of the low energy surface plasmon (ω around 2-3 eV) in simple metal clusters [10, 4]. "Volume" modes are expected to appear in the UV regime around 5-6 eV [7] and $S(\mathbf{k}, \omega)$ should as well describe these excitations finite systems.

Time Dependent Density Functional Theory (TDDFT) in real time offers a complementary way to address such questions [11]. The theory has proven robust and flexible for the analysis of cluster dynamics [12, 2], especially at the Local Density Approximation level (TDLDA). Unlike linear response theory, TDLDA is not restricted to the linear domain and allows to address a large variety of dynamical scenarios [12, 3]. It remains, of course, a mean field theory ans is, therefore, limited to moderate excitation amplitudes. In the realm of larger excitations, TDDFT should in principle be complemented by a dissipative component built on top of TDLDA. The rigorous inclusion of dissipation in finite quantum systems is still an open problem, but there exist pragmatic approaches which allow to estimate the potential impact of such effects [13]. Among others aspects it was shown that the surface plasmon response can be affected by dissipative effects. The case of volume plasmons has not yet been addressed in this way. It would nevertheless be interesting to explore that aspect in order to see to what extent a TDLDA approach, as used in [14], is validated at pure TDLDA level, which in turn would further justify other approaches as explored in [14]. This is the purpose of this short paper to explore that case.

The theoretical issues are addressed in section 2. A few illustrative results are presented in section 3, conclusions are drawn in section 4.

2. Theoretical approach

2.1. Mean field basics

Our starting point is a mean-field description of electron dynamics. We treat the systems by time-dependent density functional theory at the level of the Time-Dependent Local-Density Approximation (TDLDA). We formulate TDLDA within the usual Kohn-Sham picture on the basis of occupied single-particle (s.p.) wavefunctions $\{|\varphi_i\rangle, i = 1 \ldots N\}$ determined by the time-dependent Kohn-Sham equation

$$\mathrm{i}\partial_t|\varphi_i\rangle = \hat{h}[\varrho]|\phi_i\rangle \qquad (1)$$

At the TDLDA level, the Kohn-Sham mean-field Hamiltonian $\hat{h}[\varrho]$ is a functional of the instantaneous local density $\varrho(\mathbf{r},t) = \sum_i |\varphi_i(\mathbf{r},t)|^2$ [15].

The time propagation can be expressed in a simple matrix form in terms of the one-body density matrix $\hat{\rho}$:

$$\mathrm{i}\partial_t\hat{\rho} = \left[\hat{h}[\varrho], \hat{\rho}\right] . \qquad (2)$$

In general, $\hat{\rho}$ and $\hat{h}$ are not simultaneously diagonal during time evolution. One often prefers the basis of natural orbitals in which $\hat{\rho}$ is diagonal (and $\hat{h}$ is not). The one-body density matrix is then expressed as

$$\hat{\rho} = \sum_{\alpha=1}^{\infty} |\phi_\alpha\rangle W_\alpha \langle\phi_\alpha| \quad . \qquad (3)$$

where the weights W_α are the probability with which a state $|\phi_\alpha\rangle$ is occupied. The local density is then expressed as $\varrho(\mathbf{r},t) = \sum_\alpha W_\alpha|\phi_\alpha(\mathbf{r},t)|^2$. In a pure TDLDA evolution, the weights W_α do not depend on time and the time propagation is fully accommodated by the time evolution of the $|\phi_\alpha(t)\rangle$. Changes in W_α and fractional W_α can only be generated by dissipation which is an effect beyond TDLDA.

2.2. A quantum Relaxation Time Ansatz (RTA)

A formally simple way to include a dissipative component into mean-field evolution by is to add a collision term $\mathrm{i}\partial_t\hat{\rho} - [\hat{h}, \hat{\rho}] = \hat{I}[\hat{\rho}]$ [16, 17] to the time evolution of the density matrix Eq. (2). Following experience gained in classical and semi-classical cases [18], we approximate the quantum collision term by a Relaxation Time Approximation (RTA) which leads to the dynamical equation

$$\partial_t\hat{\rho} \;=\; -\mathrm{i}[\hat{h}, \hat{\rho}] - \frac{1}{\tau_{\mathrm{relax}}}\left(\hat{\rho} - \hat{\rho}_{\mathrm{eq}}[\varrho, \mathbf{j}, E]\right) , \qquad (4)$$

Two key quantities enter this expression: $\hat{\rho}_{\mathrm{eq}}$ represents the density operator of the thermally equilibrated state, "constrained by" the local density $\varrho(\mathbf{r},t)$, current distribution $\mathbf{j}(\mathbf{r},t)$ and total energy $E(t)$ at the given instant t where the collision term is applied; τ_{relax} is the relaxation time

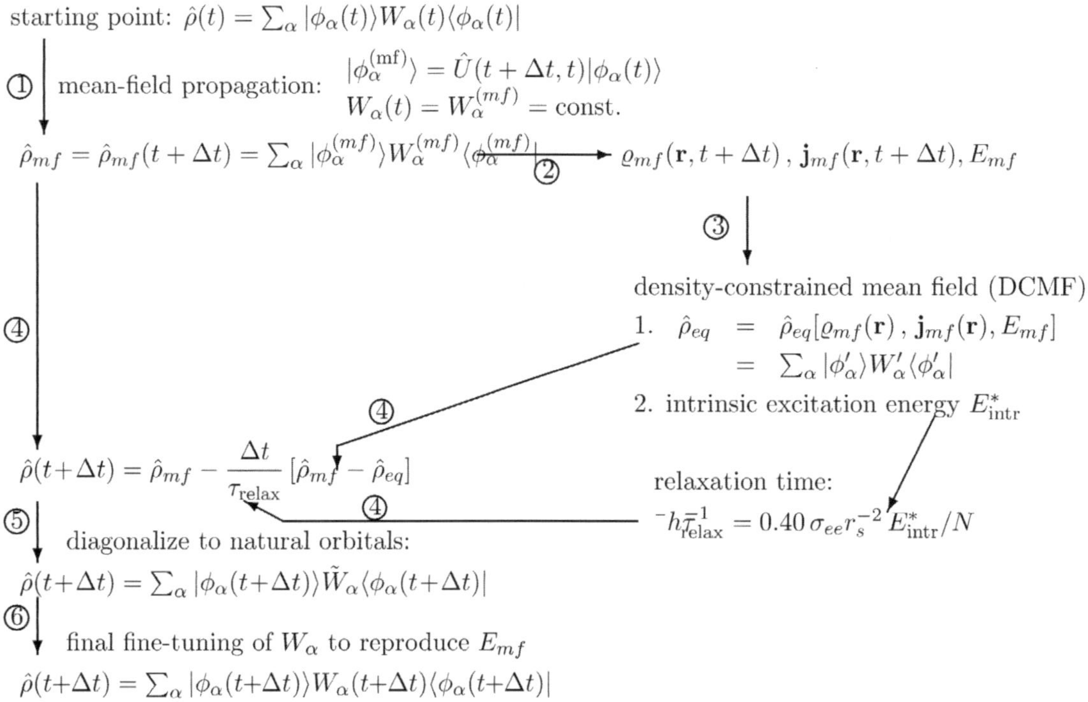

Figure 1. Sketch of the scheme for performing RTA time step $t \longrightarrow t+\Delta t$. Taken from [13].

which is estimated by semi-classical Fermi liquid theory [13]. In the Sodium clusters examined here, τ_{relax} is

$$\frac{\hbar}{\tau_{\mathrm{relax}}} = 0.40\frac{\sigma_{ee}}{r_s^2}\frac{E^*_{\mathrm{intr}}}{N}\,. \tag{5}$$

Above, we have introduced the intrinsic (thermal) energy E^*_{intr} of the system and the in-medium electron-electron cross section σ_{ee} (r_s is the effective Wigner-Seitz radius of the electron cloud). See [13] for technical details.

Note that the RTA equation (4) is very involved because many entries depend in various ways on the actual state $\hat{\rho}(t)$ of the system. Equation (4) determines the mean field Hamiltonian $\hat{h} = \hat{h}[\varrho]$ as well as the instantaneous equilibrium density $\hat{\rho}_{\mathrm{eq}}$, which is attained by calculations with with constraint on the actual $\varrho(\mathbf{r})$, $\mathbf{j}(\mathbf{r})$ and energy E. which is obtained by calculations constrained by the actual $\varrho(\mathbf{r})$, $\mathbf{j}(\mathbf{r})$ and energy E. In practice, the RTA evolution is performed every Δt time intervals on top of mean-field evolution performed at a finer δt time step. For the sake of clarity the actual RTA evolution is sketched in Figure 1.

2.3. TDLDA and RTA description of finite momentum response

The goal of this paper is to investigate the importance of dissipation for the response at finite momentum transfer. T h e response of the system can be rather simply examined within the real time TDLDA used here and similarly within RTA described above. This response has been widely studied in the long-wavelength limit. Real time TDLDA (or RTA) evolution is initialized by an initial boost $exp(ia\,z)$ acting on all single electron wavefunctions (φ_i or ϕ_α, which coincide at initial time). The boost amplitude controls the degree of linearity of the process. The response is analyzed via the dipole operator along the boost excitation axis z as $D(t) = \int d^3r z\varrho(\mathbf{r})$. Its

Fourier transform to the frequency domain $D(t) \to \tilde{D}(\omega)$ provides a direct access to the spectral information [19].

The above boost mechanism dominantly excites the surface plasmon. In order to access the volume plasmon we use a finite momentum k excitation operator :

$$\varphi_i(\mathbf{r},t) \longrightarrow \exp(\mathrm{i}aj_1(kr)z/r)\varphi_i(\mathbf{r},t) \tag{6}$$

where $r^2 = |\mathbf{r}|^2$, a n d $j_1(x)$ is the spherical Bessel function of order 1. The amplitude parameter a can again be chosen freely, for example to remain in the linear response regime where it disappears from the final analysis. The response is analyzed as in case of the surface plasmon but now considering the Bessel weighted dipole

$$D_1(t) = \int d^3r\varrho(\mathbf{r},t)j_1(kr)z/r \tag{7}$$

in time. Spectral information is then again obtained by Fourier transforming $D_1(t) \to \tilde{D}_1(\omega)$. As is obvious from the above definitions, the TDLDA and RTA frameworks are equally suited to study the response at any finite momentum: the excitation mechanism is performed the same way in both cases and the response is analyzed via observables directly attained though the matter density $\varrho(\mathbf{r})$.

2.4. *The response approach: An alternative to real time TDLDA*

An alternative procedure to calculate the response to an external field $\delta U_{\mathrm{ext}}(\mathbf{r},t)$ is to linearize equations of motion and compute the induced density $\delta\varrho(\mathbf{r},t)$ directly expressed from the density-density response function $\chi(\mathbf{r},\mathbf{r}';\omega)$:

$$\delta n(\mathbf{r},t) = n(\mathbf{r})\int d^3r'\chi(\mathbf{r},\mathbf{r}';\omega)\delta U_{\mathrm{ext}}(\mathbf{r},\omega)\,. \tag{8}$$

.

The response function can be computed starting from the stationary Kohn-Sham wave functions $\varphi_i(\mathbf{r})$ and restricting excited states to one-particle-one-hole excitations. This basically leads to the RPA response function

$$\chi(\mathbf{r},\mathbf{r}';\omega) = \left[[\mathbf{1} - \chi_0 * V_{\mathrm{p-h}}]^{-1} * \chi_0\right](\mathbf{r},\mathbf{r}';\omega) \tag{9}$$

where the asterisk and the inverse is understood as the convolution product. The χ_0 is the non interacting response function and the particle-hole interaction $V_{\mathrm{p-h}}$, LDA level, consists of the Coulomb potential and a contact interaction associated to the exchange-correlation correction,

$$V_{\mathrm{p-h}}(\mathbf{r},\mathbf{r}') = V_c(|\mathbf{r}-\mathbf{r}'|) + \frac{d^2\varrho\varepsilon_{xc}(\varrho)}{d\varrho^2}\delta(\mathbf{r}-\mathbf{r}') \tag{10}$$

where $\varepsilon_{xc}(\varrho)$ is the local exchange-correlation energy density.

Only the diagonal part of the density-density response function is probed in inclusive scattering experiments

$$\chi(\mathbf{k};\omega) = \frac{1}{N}\int d^3rd^3r'e^{i(\mathbf{k}\cdot(\mathbf{r}-\mathbf{r}'))}\chi(\mathbf{r},\mathbf{r}';\omega)\,. \tag{11}$$

and the dynamic structure function $S(\mathbf{k},\omega)$ is then

$$S(\mathbf{k},\omega) = -\frac{1}{\pi}\mathcal{I}m\chi(\mathbf{k};\omega) \qquad (\omega > 0). \tag{12}$$

Electronic and optical properties of a double quantum dot molecule with Kane's dispersion law

K G Dvoyan[1], A A Tshantshapanyan, S G Matinyan and B Vlahovic

Department of Physics, North Carolina Central University, 1801 Fayetteville St., Durham, NC, 27707, USA

E-mail: kdvoyan@nccu.edu

Abstract. In the framework of the adiabatic approximation the electronic states and direct interband absorption of light in the double quantum dot molecule (QDM) are discussed. Cases of both standard parabolic and Kane's dispersion law are considered. Analytical expressions for the wave functions and energy spectrum of the electron in the QDM are treated. The split of energy levels due to the possibility of the electron tunneling between quantum dots (QDs) in the molecule is revealed. The corresponding selection rules of quantum transitions for the direct interband absorption of light are obtained. The absorption edge behavior characteristics depending on the QDs geometrical sizes and the width of the QDs connecting region are revealed as well.

1. Introduction

QDMs are one of the modern semiconductor low-dimensional systems, which are of a great interest due to their possible applications in photonics, in the design of photovoltaic devices and various high-precision detectors, as well as the development of quantum computers [1-4]. There are many works devoted to detailed and versatile studies of vertically stacked (grown one over the other) systems of QDs. It is relatively easy to grow vertically arranged QDMs due to the internal symmetry of semiconductors. The purposeful experimental and theoretical study of horizontally stacked QDMs remains a significant challenge for the research groups [5-10]. However, during the growth of real semiconductor structures, in addition to the separately arranged QDs, both symmetric and asymmetric QDMs arise inevitably. Such original "errors" in the experiments lead to useful results, as history of science shows. The tunneling of charge carriers (CCs) from one QD to another becomes possible in closely located QDs. The splitting of the energy levels in the QDM, due to possible tunneling, allows one to consider these objects as an artificial analogue of the molecular orbitals of real molecules. Similarity of the electron orbitals of different analyte molecules to the levels of the QDM also allows the tunneling of electrons between the analyte molecules and the QDM. Practical implementation of this effect opens wide possibilities for the applications of QDMs for the design of various biochemical sensors and detectors, which may potentially lead to label free detection and identification of a wide range of analytes with a single molecule sensitivity.

[1] Corresponding author: K. G. Dvoyan, E-mail: kdvoyan@nccu.edu

Due to the interaction between, the problem of finding of electronic states and wave functions (WFs) in the QDM is much more complicated. At the same time, their properties strongly depend on the external form, confining potential, inhomogeneity of CCs effective masses, and the number of QDs in the molecule. In order to understand QDMs properties, one needs to understand the essence of their correlations and coupling between them. Therefore, it is very important to develop clear criteria for classification and identification of QDs correlations and QDMs formations. Despite the objective difficulties, along with the correct choice of the confining potential, it is also very important task to describe an external form of the molecule as a whole. First proposed method for describing such systems are based on the following: the properties of individual QDs of different shapes are considered, and depending on their relative position (close or far) corresponding corrections due to their interaction are introduced [6]. In other words, several correlation terms are added in the Hamiltonian, which describes the interactions of the separately grown, but closely arranged QDs. However, most often there is no clear boundary between QDs in the molecule in real grown structures and a binding region – an isthmus – appears between QDs. Obviously, in such systems above mentioned description method will not be accurate. It is well known that even a small change in external shape of QDs leads to a significant change of the CC energy spectrum, and consequently to other physical properties of the sample. From this perspective, revealing a correlation between QDs, reduced to an accurate description of their binding region (isthmus), and a detailed study and estimation of the contribution of its presence on the behavior of the CC within the system. We proposed several models describing the correlation between QDs in molecules, with a single confining potential [7]. In particular, the additional splitting of the electronic levels due to the tunneling in presence of the central QDs was revealed. In the paper [11], authors considered the electronic structure of two laterally coupled Gaussian QDs filled with two particles. Their research shows that such structures have highly modifiable properties, promoting it as an interesting quantum device, showing the possible use of these states as a quantum bit gate.

a)

b)

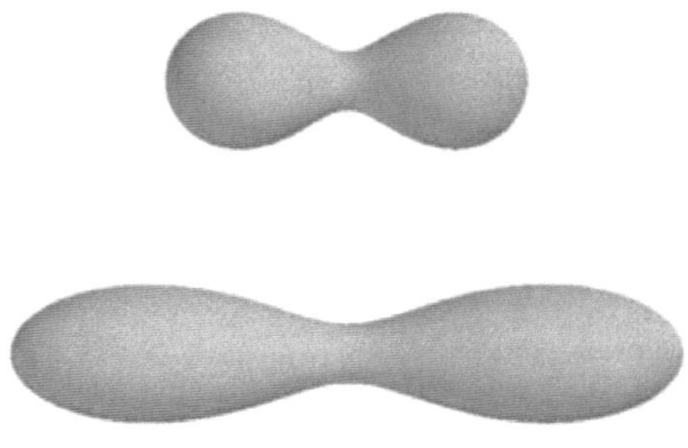

Figure 1. Symmetric QDM a) Cassini lemniscate shaped, b) ellipsoidal Cassini lemniscate shaped.

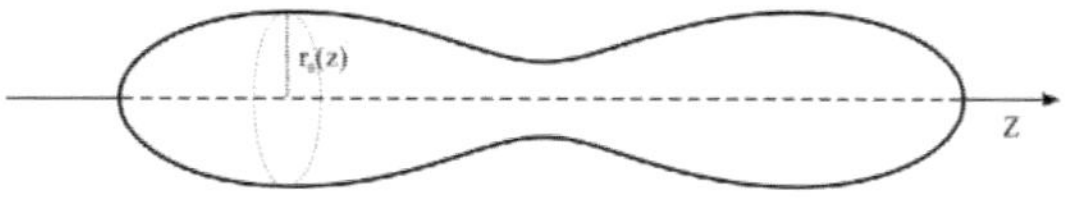

Figure 2. QDM cross-section.

Another major factor influencing the properties of the CC in the low-dimensional systems is the internal structure of matter composing the grown system. It is known that in the narrow-gap semiconductors the CCs' dispersion law is completely different from the standard parabolic law. However, there are many papers in which more complicated dependence of the CC effective mass on the energy is considered [12-16] in the framework of the Kane's theory. For example, for the narrow-gap QDs of InSb, the dispersion law of CCs is non-parabolic and it is well described by the Kane's two-band mirror model [14,17]. Within the framework of the two-band approximation the electron dispersion law (light hole) formally coincides with the relativistic law. For more successful and realistic approximation of real grown double QDs, in the present paper, the electronic states and direct interband absorption of light in the QDM having the shape of the Cassini lemniscate of revolution for the standard and Kane's dispersion laws are calculated. The potential applications of these systems include sensors and detectors for various optoelectronic, medical and IT applications.

2. Electronic States

Consider an impermeable symmetric QDM consisting of two QDs having a shape of Cassini lemniscate of revolution (see figure 1). Then the potential energy of the particle in cylindrical coordinates can be written as

$$U=\begin{cases}0,\left(\rho^2+\lambda Z^2\right)^2-2c_1^2\left(-\rho^2+\lambda Z^2\right)-a_1^4+c_1^4\le 0\\ \infty,\left(\rho^2+\lambda Z^2\right)^2-2c_1^2\left(-\rho^2+\lambda Z^2\right)-a_1^4+c_1^4>0\end{cases}, \qquad (1)$$

where c_1 is a focal length of the lemniscate of revolution, a_1 is a product of distances from foci to any point on the surface, and λ is a parameter of ellipsoidality (prolateness) of QDs. Here we will discuss the case of equal effective masses in both QDs $m_{e_1}^*=m_{e_2}^*=m_e^*$, however, one can easily take into account their difference depending on the geometrical directions (for example, along the axis of the molecule and perpendicular to it $m_{e\perp}^*\neq m_{e\parallel}^*$). Obtained results can be generalized to the case of the complete difference between the effective masses.

2.1. Parabolic dispersion law

In the regime of the strong size quantization (SQ), when the condition $2c_1 \ll a_B$ (a_B is the effective Bohr radius) takes place, the energy of the Coulomb interaction between an electron and a hole is much less than the energy caused by the SQ contribution. In this approximation the Coulomb interaction between the electron and hole can be neglected, and the problem reduces to the determination of their energy states separately. First, we will discuss the case of the standard parabolic dispersion law of charge carriers. Then the Hamiltonian in cylindrical coordinates can be written

$$\hat{\hat{H}}=-\frac{\hbar^2}{2m_{e\perp}^*}\left(\frac{\partial^2}{\partial\rho^2}+\frac{1}{\rho}\frac{\partial}{\partial\rho}+\frac{1}{\rho^2}\frac{\partial^2}{\partial\varphi^2}\right)-\frac{\hbar^2}{2m_{e\parallel}^*}\frac{\partial^2}{\partial Z^2}+U\left(\rho,\varphi,Z\right). \qquad (2)$$

It follows from the geometrical shape of the QDM that the particle motion in the OZ direction is faster than in the perpendicular direction. This fact allows one to apply the geometric adiabatic approximation. The system's Hamiltonian can be represented as a sum of the "fast" $\hat{H}_1$ and "slow" $\hat{H}_1$ subsystems Hamiltonians in dimensionless units: $\hat{H}=\hat{H}_1+\hat{H}_2+U\left(r,\varphi,z\right)$, where

$$\hat{H}_1=-\frac{1}{4}\left(\frac{\partial^2}{\partial r^2}+\frac{1}{r}\frac{\partial}{\partial r}+\frac{1}{r^2}\frac{\partial^2}{\partial\varphi^2}\right),\quad \hat{H}_2=-\frac{\beta}{4}\frac{\partial^2}{\partial z^2}, \qquad (3)$$

and the following notations are introduced: $\hat{H}=\frac{\hat{\hat{H}}}{E_R}$, $r=\frac{\rho}{a_B}$, $z=\frac{Z}{a_B}$, $\beta=\frac{m_{e\perp}^*}{m_{e\parallel}^*}$, $E_R=\frac{\hbar^2}{2m_e^*a_B^2}$ is the effective Rydberg energy of an electron, $a_B=\frac{\kappa\hbar^2}{m_e^*e^2}$ is the effective Bohr radius of an electron, κ is a dielectric permittivity, e and m_e^* are the charge and effective mass of an electron, respectively. WFs of the problem is sought in the form

$$\Psi\left(r,\varphi,z\right)=Ce^{im\varphi}R\left(r;z\right)\chi\left(z\right), \qquad (4)$$

where C is a normalization constant. At a fixed value of the coordinate z of the "slow" subsystem the particle motion is localized in a two-dimensional potential well with an effective width (figure 2)

$$r_0\left(z\right)=\sqrt{\sqrt{4\lambda c^2z^2+a^4}-\lambda z^2-c^2}, \qquad (5)$$

where $a=\frac{a_1}{a_B}$, $c=\frac{c_1}{a_B}$. After simple transformations, the Schrödinger equation

$$-\frac{1}{4}\left(\frac{\partial^2}{\partial r^2}+\frac{1}{r}\frac{\partial}{\partial r}+\frac{1}{r^2}\frac{\partial^2}{\partial\varphi^2}\right)R(r;z)=\varepsilon(z)R(r;z), \tag{6}$$

reduces to the following Bessel equation

$$R''(\eta,z)+\frac{1}{\eta}R'(\eta,z)+\left(1-\frac{m^2}{\eta^2}\right)R(\eta,z)=0, \tag{7}$$

where $\eta=2\sqrt{\varepsilon(z)}\,r$. The solutions of the equation (7) are the first kind Bessel functions:

$$R(r;z)=A(z)J_m\left(\sqrt{\varepsilon(z)}\,r\right), \tag{8}$$

where $A(z)$ is a normalization constant. From the boundary conditions one obtains the following expression for the "effective" two-dimensional motion energy:

$$\varepsilon(z)=\frac{\alpha_{n,m}^2}{4r_0^2(z)}=\frac{\alpha_{n,m}^2}{4\left(\sqrt{4\lambda c^2z^2+a^4}-\lambda z^2-c^2\right)}, \tag{9}$$

where $\alpha_{n,m}$ are roots of the Bessel functions. The energy (9) plays role of an effective potential energy in the Schrödinger equation of the "slow" subsystem, which obviously cannot be solved exactly. However, for the lower levels of the energy spectrum, the particle is localized in the geometrical center of one of the QDs, with coordinates $z_0=\pm\frac{\sqrt{4c^4-a^4}}{2c\sqrt{\lambda}}$. After expanding (9) in a power series around these points one gets the expression (figure 3)

$$\varepsilon(z)\simeq\frac{\alpha_{n,m}^2c^2}{a^4}+\frac{\lambda\alpha_{n,m}^2\left(4c^4-a^4\right)}{a^8}\left(z\pm\frac{\sqrt{4c^4-a^4}}{2c\sqrt{\lambda}}\right)^2, \tag{10}$$

or

$$\varepsilon(z)\simeq\varepsilon_0+\gamma^2\frac{(z\pm z_0)^2}{16}, \tag{11}$$

with $\varepsilon_0=\frac{4\alpha_{n,m}^2c^2}{a^4}$ and $\gamma=\frac{4\alpha_{n,m}\sqrt{\lambda\left(4c^4-a^4\right)}}{a^4}$.

Note, that in the vast majority cases of the adiabatic approximation applications, in contrast to this specific one, the effective potential of the "slow" subsystem turns out parabolic. Here the electron motion in the direction of the "slow" subsystem occurs under the influence of the double parabolic effective potential as a consequence of the specific geometric form of the QDM. Further, following the geometric adiabatic approximation technique, it is necessary to solve the Schrödinger equation with the effective potential (11):

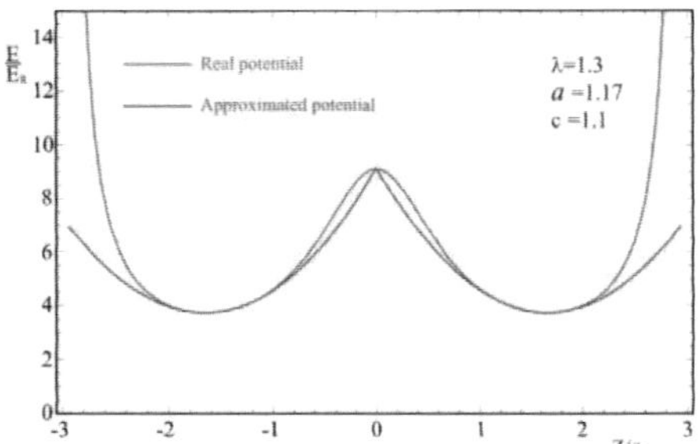

Figure 3. Confinement potential of the "slow" subsystem.

$$\left(-\frac{\beta}{4}\frac{\partial^2}{\partial z^2}+\varepsilon_0+\frac{\gamma^2(z\pm z_0)^2}{16}\right)\chi(z)=\varepsilon\,\chi(z). \tag{12}$$

After the change of variables $\xi^{\pm}=\sqrt{\frac{\gamma}{\sqrt{\beta}}}(z\pm z_0)$, one obtains the parabolic cylinder equation for the left (ξ^-) and right (ξ^+) sides of the double parabolic potential, respectively:

$$\left(\frac{\partial^2}{\partial\xi^2}+\delta-\frac{\xi^2}{4}\right)\chi(\xi)=0, \tag{13}$$

where $\delta=\frac{4(\varepsilon-\varepsilon_0)}{\gamma\sqrt{\beta}}$. The solutions of the equation (13) are the parabolic cylinder functions $D_\nu(t)$:

$$\chi^{\pm}(z)=C^{\pm}D_\nu\left(\sqrt{\frac{\gamma}{\sqrt{\beta}}}(z\pm z_0)\right), \tag{14}$$

where $\nu=\delta-\frac{1}{2}=\frac{4(\varepsilon-\varepsilon_0)}{\gamma\sqrt{\beta}}-\frac{1}{2}$. Finally, in the strong SQ regime the WFs of the electron in the QDM can be written as

$$\Psi(r,\varphi,z)=Ce^{im\varphi}J_m\left(\frac{\alpha_{n,m}}{r_0(z)}r\right)D_\nu\left(\sqrt{\frac{\gamma}{\sqrt{\beta}}}(z\pm z_0)\right). \tag{15}$$

The full energy of the electron in the strong SQ regime in the QDM with the standard parabolic dispersion law is determined from the sewing of WFs (14) at the point $z=0$:

$$\left.\frac{\chi^-(z)'}{\chi^-(z)}\right|_{z=0}=\left.\frac{\chi^+(z)'}{\chi^+(z)}\right|_{z=0}. \tag{16}$$

2.2 Kane's dispersion law

Let us consider the QDM with Kane's dispersion law $E^2=P^2S^2+m_e^{*2}S^4$, where $S\sim10^8\,cm/\sec$ is the parameter related to the semiconductor bandgap $E_g=2m_e^*S^2$. One can write the Klein-Gordon equation [18] for a symmetric QDM in the strong SQ regime, when Coulomb interaction between the electron and the hole is neglected:

$$\left(P^2S^2+m_e^{*2}S^4\right)\Psi(\rho,\varphi,Z)=\left(E-U(\rho,\varphi,Z)\right)^2\Psi(\rho,\varphi,Z). \tag{17}$$

After simple transformations, the (17) can be written as the reduced Schrödinger equation

$$\frac{P^2}{2m_e^*}\Psi(\rho,\varphi,Z)=E_1\Psi(\rho,\varphi,Z), \tag{18}$$

where $E_1=\frac{E^2-m_e^{*2}S^4}{2m_e^{*2}S^2}=\frac{4E^2-E_g^2}{4E_g}$. Then, repeating the above procedure of the adiabatic approximation, one gets the following expression for the OZ direction WFs:

$$\chi_{\pm}(z)=C_{\pm}D_{\nu_K}\left(\sqrt{\gamma}(z\pm z_0)\right), \tag{19}$$

where $\nu_K=\frac{4(\varepsilon_1-\varepsilon_0)}{\gamma}-\frac{1}{2}=\frac{4\left(\frac{4\varepsilon^2-\varepsilon_g^2}{4\varepsilon_g}-\varepsilon_0\right)}{\gamma}-\frac{1}{2}$. Note, that in contrast to the parabolic dispersion law, the WFs dependence on the CC energy is non-linear in the case of the Kane's dispersion law. Finally, the WFs of the electron in the QDM with Kane's dispersion law can be written as

$$\Psi(r,\varphi,z)=Ce^{im\varphi}J_m\left(\frac{\alpha_{n,m}}{r_0(z)}r\right)D_{\nu_K}\left(\sqrt{\frac{\gamma}{\sqrt{\beta}}}(z\pm z_0)\right). \tag{20}$$

The total energy of the electron in the *InSb* QDM in the strong SQ regime is determined from the sewing of WFs (19) at the point $z=0$ (see (16)).

3. Direct interband absorption of light

Consider the direct interband absorption of light in the QDM in the strong SQ regime, when the Coulomb interaction between the electron and a hole is neglected. The case of a heavy hole is discussed ($m_e^* \ll m_h^*$), and the absorption coefficient is determined by the expression [19]

$$K = A\sum_{\alpha,\alpha'}\left|\int \Psi_\alpha^e \Psi_{\alpha'}^h d\vec{r}\right|^2 \delta\left(\hbar\Omega - E_g - E_\alpha^e - E_{\alpha'}^h\right), \tag{21}$$

where α and α' are the quantum number (QN) sets corresponding to the electron and a heavy hole, E_g is the bandgap of a bulk semiconductor, Ω is an incident light frequency, A is a quantity proportional to the square of the matrix element taken by the Bloch functions. After simple calculations one gets the following expression for the absorption edge (AE) W_{100}:

$$W_{100} = 1 + \varepsilon\frac{d^2}{a_B^2}, \tag{22}$$

where $W_{100} = \frac{\hbar\Omega_{100}}{E_g}$, $d = \frac{\hbar}{\sqrt{2\mu E_g}}$, and $\mu = \frac{m_e^* m_h^*}{m_e^* + m_h^*}$ is the reduced mass of the exciton. The formula (22) describes the dependence of the effective bandgap on the parameters a_1 and c_1 (from the expression of the energy). With increasing the parameter a_1, the AE shifts to the long wave region, whereas the increase in the c_1 shifts AE to the short wave region.

Consider now the selection rules in quantum transitions. Quantum transitions for the energy levels allowed for the magnetic QNs $m = -m'$, and for the "fast" subsystem QNs $n = n'$. The continuity of logarithmic derivatives of the WFs (16) leads to the complete removal of the "slow" subsystem QNs selection rules. Note, that the analytical expression (22) is given taking into account above mentioned selection rules.

4. Discussion

When the geometric adiabatic approximation is applied, there is usually a family of the "slow" subsystem energy sublevels for each energy level of the "fast" subsystem. In most cases, the "slow" subsystem's energy levels turn equidistant, since the effective confinement potential is parabolic. As it is seen from the (11), the confinement potential of the "slow" subsystem has the form of a double parabola, in contrast to the other cases of the geometric adiabatic approximation application. Obviously, the family of the "slow" subsystem energy levels are not equidistant due to the formation of that kind of the confinement potential. Moreover, levels are double split due to the tunneling

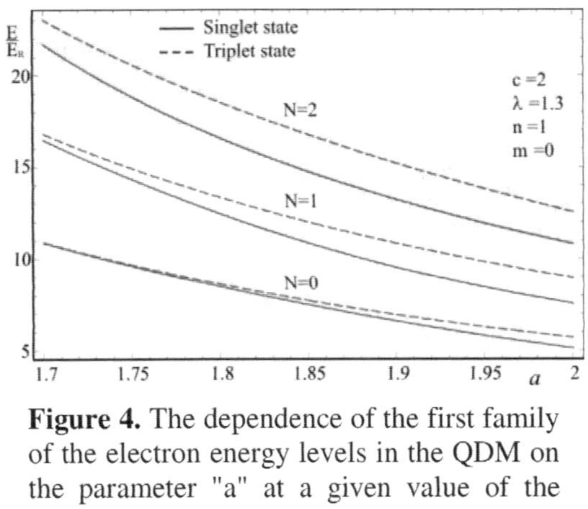

Figure 4. The dependence of the first family of the electron energy levels in the QDM on the parameter "a" at a given value of the lemniscate focal length "c".

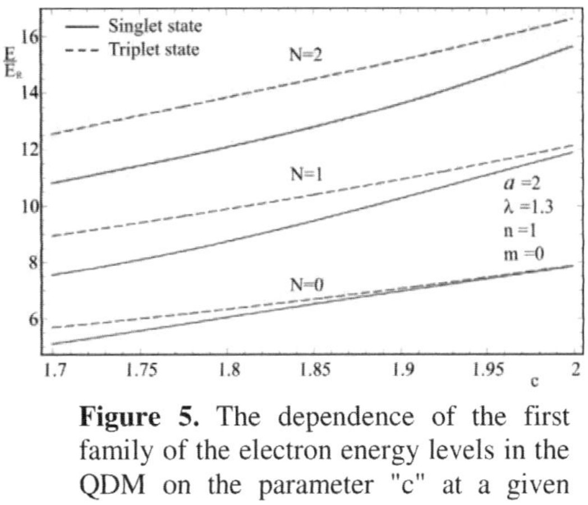

Figure 5. The dependence of the first family of the electron energy levels in the QDM on the parameter "c" at a given value of the parameter "a".

possibility between two QDs. In other words, split levels of the "slow" subsystem positioned on each energy level of the "fast" subsystem. Note that the magnitude of the splitting depends on the geometrical parameters of the QDM and the "fast" subsystem QN as well. The first circumstance is conditioned by the electron probability of tunneling between quantum wells, and the latter one is the consequence of the external shape of the QDM, which allows the application of the adiabatic approximation.

Figure 4 represents the dependence of the first family of the energy levels in the QDM on the parameter a at a given value of the lemniscate focal length c. As it is seen from the figure, with the increase in the parameter a, the energy curves decrease. This is a consequence of the QDM geometrical sizes increasing, which is in turn a result of the increase in the isthmus connecting two QDs. At the same time, the magnitude of the splitting of levels increases. This is due to the fact that the probability of the electron tunneling from one QD to another increases for large values of a. It should be noted that for small values of the width of the isthmus, the tunneling probability decreases, curves of the singlet and triplet states merge, and the distance between them becomes equidistant. In other words, in the case of weakly interacting QDs (low tunneling probability), the equidistant

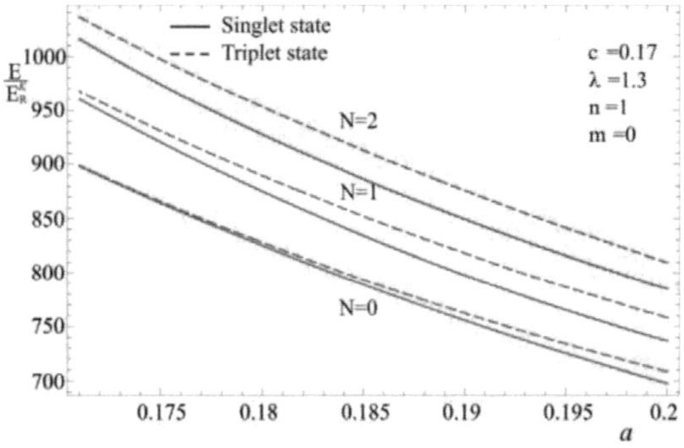

Figure 6. The dependence of the first family of the electron energy levels in the *InSb* QDM on the parameter "a" at a given value of the lemniscate focal length "c".

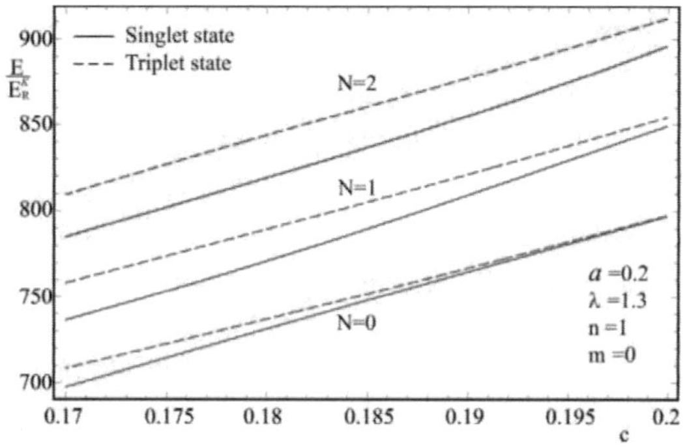

Figure 7. The dependence of the first family of the electron energy levels in the *InSb* QDM on the parameter "c" at a given value of the parameter "a".

character of the energy sublevels' families restores. The tunneling probability increases with increasing energy, as expected. The splitting magnitude is more for the higher positioned levels. Thus, at $a=2$ and $N=0$, the singlet-triplet energy levels splitting is $\Delta E \approx 0.35E_R$, whereas at $N=1$ the splitting is $\Delta E \approx 1.22E_R$. Numerical calculations for the parabolic dispersion law are performed for the *GaAs* QDM with the following parameters: $m_e^* = 0.067m_e$, $m_e^* = 0.12m_h^*$, where m_e is a free electron mass, $\kappa = 13.8$, $E_R = 5.275meV$, $a_B = 104A^{\circ}$ and $a_h = 15A^{\circ}$ are the effective Bohr radii for the electron and a hole, respectively, $E_g = 1.43eV$ is a bandgap of a bulk semiconductor. Opposite behavior of the energy levels is observed in Fig. 5., which shows the dependence of the first family of the electron energy levels in the QDM on the lemniscate focal length c at a given value of the parameter a. Thus, increasing the parameter c leads to the increase in the energy levels, conditioned by the increase in the distance between two QDs and narrowing of the isthmus connecting them. At the same time, the increase in the lemniscate focal length also leads to the decrease in the electron tunneling probability. As a consequence, the magnitude of the splitting of the energy levels decrease. Note, that at the value $a=c$, the Cassini lemniscate becomes the Bernoulli lemniscate, which corresponds to the case of zero width isthmus between closely spaced QDs. However, the absence of the isthmus does not result in a complete removal of the splitting. Thus, at the value of $a=c=2$ the excited levels are still split. This is due to a non-zero probability of the electron tunneling. Further increasing of the focal parameter leads to the final isolation of the QD. Note, that just in this case there is a recovery of the equidistant energy levels of the individual ellipsoidal QDs [20]. It is also important to note that the increase in the splitting of levels with the QN increasing, occurs only for the symmetric QDM. When the symmetry is violated, the picture changes dramatically. In the case of non-identical QDs, the energy "zone" is formed for the probable tunneling, which position depends on the difference of sizes of QDs and the QDM (see e. g. [21]).

Numerical calculations in the case of the Kane's dispersion law are performed for the *InSb* QDM with the following parameters: $m_e^* = 0.013m_e$, $m_e^* \approx m_h^*$, where m_e is a free electron mass, $E_R^K = 6 \cdot 10^{-4} eV$, $a_B^K = 500A^\circ$ is the electron effective Bohr radius, $E_g^K = 0.23eV$ is a bandgap of a bulk semiconductor. As it is seen, the strong SQ regime is achievable much easier for the semiconductor with the Kane's dispersion law, due to the small effective mass of the electron and the big value of effective Bohr radius.

Figure 6 illustrates the dependence of the first family of the electron energy levels in the QDM with the Kane's dispersion law on the parameter a at the given value of the focal parameter c. As can be seen, with an increase in the parameter a, the splitting of levels increases. However, due to the dispersion law, the magnitude of the splitting is different. Thus, at $a = 0.2$ and $c = 0.17$ the splitting is $\Delta E^K \approx 11.32E_R^K$, or $\Delta E^K \approx 6.79meV$, whereas for the parabolic dispersion law one gets $\Delta E \approx 1.846meV$. This significant difference is due to the fact that in the strong SQ regime the difference between CC becomes one of the key factors affecting the motion of the particle [16]. The similar dependence of the energy on the focal parameter c at the given value of the parameter a is shown on the figure 7. Here, with an increase in the parameter c, the splitting of levels decreases, as expected.

Figure 8 plots the dependence of the edge frequencies of the direct interband absorption of light in the QDM versus the width of the isthmus between two QDs in the strong SQ regime. As it is

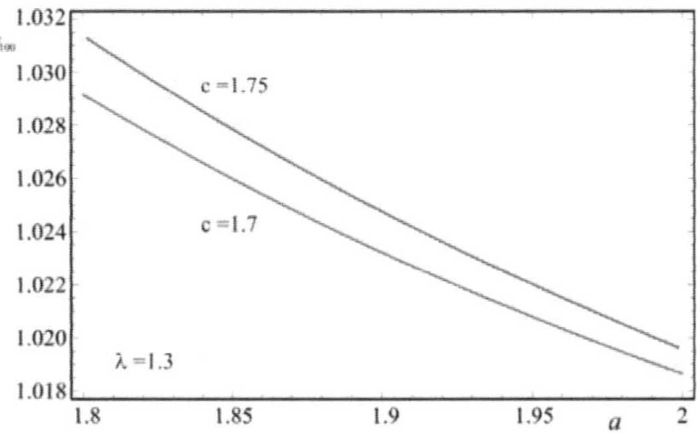

Figure 8. The dependence of the AE on the parameter "a" at the given values of the parameter "c".

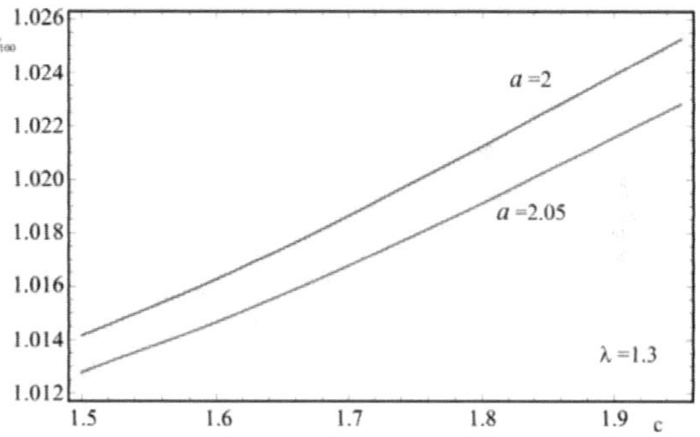

Figure 9. The dependence of the AE on the parameter "c" at the given values of the parameter "a".

seen from the figure, the increase in the isthmus width results in the decrease in the AE due to the reduction of the SQ effect (redshift). The curve corresponding to the small value of the parameter c is positioned lower, since in this case the electron localization area in the QDM is wider. The opposite pattern is observed in the figure 9, showing the dependence of the AE on the focal length distance c. As expected, the increase in c results in the blueshift of the edge frequencies, due to the SQ effect impact increase. For this reason, the curve corresponding to the small value of the a is positioned higher. Note also, that at the small values of the focal length, the AE curves are positioned closer for the different values of the parameter a. The increase in the parameter c leads to the increase in the curves discrepancy. This is due to the fact that for small values of the focal length, the width of the isthmus feels weaker due to the closely positioned QDs. At the big values of the c, the width of the isthmus becomes more influential on the motion and tunneling of the electron in the QDM.

5. Conclusion

During the growth of QDM nanostructures, it is possible to control the value of the energy of the electron and the splitting magnitude of the energy levels by manipulating sizes and the distance between QDs. The latter may promote the design of a new generation of the highly sensitive and selective sensors. In case of the coincidence of the split levels of the QDM and potentially detectable molecules of the gaseous medium, it is possible to achieve a successful tunneling of electrons from the gas molecules to the QDM. It is important also, that this tunneling is possible (most probable) only in the case of the full concurrence of interlevel distances of the QDM and a detectable molecule. This ensures high selectivity of the device, which will not fix another molecule by mistake. For the design of high-precision devices the most accurate measurements of the splitting of the energy levels are needed. For this purpose, the measurement of the light absorption frequencies in the QDM allows to establish the differences between the energy levels and the amount of the splitting.

Acknowledgements

This work is supported by the NSF (HRD-1345219) and NASA (NNX09AV07A) and in part by the M. Hildred Blewett Fellowship of the American Physical Society, www.aps.org.

References

[1] Thongkamkoon N, Patanasemakul N, Siripitakchai N, Thainoi S, Panyakeow S and Kanjanachuchai S 2011 Bimodal optical characteristics of lateral InGaAs quantum dot molecules *Journ of Crystal Growth* **324** 206-210
[2] Long B and Rong Zh 2011 Tunable spin current through a t-shaped double quantum dot molecule. *Superlattices Microstruct* **49** 543-548
[3] Boyer A, Sköld N, Stevenson R M, Farrer I, Ritchie D A and Shields A J 2011 Exciton-spin memory with a semiconductor quantum dot molecule *Phys Rev Lett* **106** 216802
[4] Rolon J E and Ulloa S E 2011 Coherent control of indirect excitonic qubits in optically driven quantum dot molecules *Phys Rev B* **82** 115307
[5] Boyer A, Sköld N, Farrer I, Ritchie D A and Shields A J 2011 Excitonic couplings and stark effect in individual quantum dot molecules *J Appl Phys* **110** 083511
[6] Zhou X, Sanwlani S, Liu W, Lee J H, Wang Zh M, Salamo G and Doty M F 2011 Spectroscopic signatures of many-body interactions and delocalized states in self-assembled lateral quantum dot molecules *Phys Rev B* **84** 205411
[7] Dvoyan K G, Kazaryan E M, Tshantshapanyan A A, Wang Zh M and Salamo G J 2011 Electronic states and light absorption in quantum dot molecule *Appl Phys Lett* **98** 203109
[8] Filikhin I, Matinyan S, Nimmo J and Vlahovic B 2011 *Physica E* **43** 1669
[9] Filikhin I, Suslov V M and Vlahovic B 2006 *Physica E* **33** 349
[10] Filikhin I, Deyneka E and Vlahovic B 2006 *Physica E* **31** 99
[11] Garcia J Z, Pietilainen P, Chen H and Chakraborty T 2008 *Physica E* **40** 2839
[12] Kane E O 1957 *J Phys Chem Solids* **1** 249
[13] Atoyan M S, Kazaryan E M, Poghosyan B Zh and Sarkisyan H A 2011 *Physica E* **43** 1592
[14] Poghosyan B Zh 2007 *Nanoscale Res Lett* **2** 515
[15] Askerov B 1985 *Electronic and Transport Phenomena in Semiconductors* (Moscow: Nauka)
[16] Avetisyan A A, Djotyan A P, Kazaryan E M and Poghosyan B G 2001 *Phys Status Solidi B* **225** 423-431
[17] Avetisyan A A, Djotyan A P, Kazaryan E M and Poghosyan B Zh 2000 *Phys Status Solidi B* **218** 441
[18] Galitsky V M, Karnakov B M and Kogan V I 1981 *Practical Quantum Mechanic* (Moscow: Nauka)
[19] Efros Al L and Efros A L 1982 *Sov Phys Semicond* **16** 772
[20] Dvoyan K G, Hayrapetyan D B and Kazaryan E M 2009 *Nanoscale Res Lett* **4** 106

[21] Vlahovic B and Dvoyan K 2015 *Nanotechnology in the Security Systems* (NATO Science for Peace and Security Series C: Environmental Security) pp 137-150

Journal of Physics: Conference Series **702** (2016) 012011 doi:10.1088/1742-6596/702/1/012011

Theory of electronic states and direct light absorption in the biconical (superellipsoidal) quantum dot

A A Tshantshapanyan, K G Dvoyan and B Vlahovic
Department of Mathematics and Physics, North Carolina Central University, 1801 Fayetteville St., Durham, NC 27707, USA

E-mail: atshants@nccu.edu

Absract. The electronic states in the *GaAs* biconical (superellipsoidal) quantum dot (QD) in a strong and weak size quantization (SQ) regimes are theoretically investigated within the framework of the geometrical adiabatic approximation. An atypical linear term in the effective confining potential forms a family of non-equidistant sublevels in the electron energy spectrum, whereas the quadratic term leads to the appearance of the equidistant energy levels as a consequence of the symmetry of the QD. In the weak size quantization regime the motion of the exciton's center-of-gravity is quantized, which leads to the appearance of the additional Coulomb sub-levels in the energy spectrum of the QD. The direct interband absorption of light in the biconical QD is also considered in both strong and weak SQ regimes. The dependences of the absorption edge on QD geometrical parameters are revealed in the linear and quadratic approximations of the effective potential adiabatic terms. Corresponding selection rules for quantum transitions are obtained as well.

1. Introduction

Modern technologies allow one to grow QDs of various shapes and sizes [1,2]. The distinctive feature of these quasi-zero dimensional structures is a linear energy spectrum of the charge carriers (CC) in them similar to atoms, which is a consequence of the full quantization of CCs motion. In recent years, there are many theoretical and experimental works, where QDs having ellipsoidal symmetry are considered [3-12]. In particular, the physical properties of the CCs in both QDs having spherical shape (ellipsoid of revolution which differs little from a shere) and in the strongly flattened and elongated ellipsoidal QDs are studied. For the first time a quantum-mechanical problem of finding the CCs energy states in the slightly flattened (elongated) ellipsoidal quantum well was formulated and solved in the first-order perturbation theory by Migdal for the droplet model of the atomic nuclei [13]. Now the solution of this problem is already used for real QDs having spherical shape.

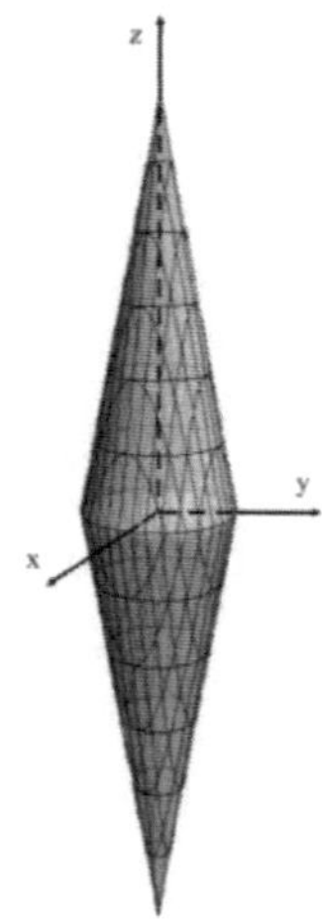

Figure 1. Biconical QD.

An exceptional result of the CCs energy spectrum in the QD with weakly expressed ellipsoidality is that the correction to the energy for different values of the quantum numbers (QNs) have different signs, which reconstruct a spectrum in a way that does not shift the center of gravity of the multiplet. It is known that even a small change in the external shape or size of the QD leads to the noticeable reconstruction of the CCs energy spectrum. Obviously, significant change in the geometric dimensions of the nanostructure leads to a fundamental change in the CCs spectrum. Thus, the effective confining parabolic potential may be formed due to the peculiarities of the QD's external form, e.g. QDs having the shape of strongly flattened (elongated) ellipsoid of revolution [14,15]. A similar situation is observed in thin lens shaped QDs [16-18]. It should be noted that the efficient parabolic potential forms due to the geometrical adiabatic approximation application. For instance, in a thin lens shaped QD having a form of a segment of a sphere or an ellipsoid, the CC motion in the direction normal to the segment is much faster than in the plane of the base; this type of separation of the CC motion into the "fast" and "slow" parts is the essence of the geometric adiabatic approximation. As a result of the strong differences in size of QD along different geometrical directions, the family of equidistant sublevels appear above the levels of the "fast" subsystem in the energy spectrum of a QD. It should be noted also that in contrast to the Born-Oppenheimer approximation, in the case of the geometric adiabatic approximation the mass of a particle does not play a significant role. However, the symmetry of a nanosructure is also important. Thus, in the case of a strongly flattened (elongated) ellipsoidal QD, the energy dependence on the geometric parameters is significantly different than in the case of the thin lens shaped QD, although the geometrical adiabatic approximation is applicable in both cases. From this point of view, reducing the degree of symmetry of the external shape of the object leads to more complex dependence on the QD sizes of the particle energy. Therefore, the investigation of strongly elongated QD having superellipsoidal symmetry is important, in particular, QDs having biconical shape. On the other hand, the formation of new size quantized (SQ) sublevels significantly affect the character of the optical absorption in semiconductor structures, which is a powerful tool for determining many characteristics of these systems: bandgaps, effective masses of electrons and holes, their mobility, dielectric constants, etc. It is expected that even at the absence of external quantizing fields, due to the geometrical reconstruction of the CC spectrum, there should appear new quantum transitions, which are absent in spherical QDs. In turn, manipulating the sizes of QDs during their growth it is possible to achieve the implementation of two SQ regimes: strong and weak. The appearance of the excitonic effects should be also taken into account in the optical absorption spectra.

2. Electronic states

Consider a superellipsoidal QD having a biconical shape (figure 1). Then the potential energy of the CC (an electron, hole, exciton) can be written as

$$U(x,y,z)=\begin{cases} 0, & \left|\sqrt{\dfrac{x^2+y^2}{a_1^2}}\right|+\left|\dfrac{z}{c_1}\right|\le 1 \\ \infty, & \left|\sqrt{\dfrac{x^2+y^2}{a_1^2}}\right|+\left|\dfrac{z}{c_1}\right|\ge 1 \end{cases}, \quad a_1 \ll c_1, \tag{1}$$

where a_1 and c_1 are a minor and major semiaxes of the bicone, correspondingly.

2.1 Strong SQ regime

First we solve the problem in the strong SQ regime, when the condition $a_{ex} \gg c_1$ takes place, a_{ex} is the effective Bohr radius. In this approximation the Coulomb interaction between an electron and hole is much less than the energy caused by the SQ contribution, and the Coulomb interaction between the electron and hole can be neglected. Then the problem reduces to the determination of CCs energy states separately. The system's Hamiltonian in cylindrical coordinates can be written as

$$\hat{\mathrm{H}}=-\frac{\hbar^2}{2m_e^*}\left(\frac{\partial^2}{\partial\rho^2}+\frac{1}{\rho}\frac{\partial}{\partial\rho}+\frac{1}{\rho^2}\frac{\partial^2}{\partial\varphi^2}+\frac{\partial^2}{\partial Z^2}\right)+U(\rho,\varphi,Z). \tag{2}$$

From the elongated geometrical form of the QD ($a_1 \ll c_1$) follows that the particle motion in the radial direction is faster than in the OZ direction. On this basis the system's Hamiltonian in dimensionless units can be represented as a sum of the "fast" $\hat{H}_1$ and "slow" $\hat{H}_2$ subsystems:

$$\hat{H}=\hat{H}_1+\hat{H}_2+U(r,\varphi,z), \tag{3}$$

where

$$\hat{H}_1 = -\left(\frac{\partial^2}{\partial r^2} + \frac{1}{r}\frac{\partial}{\partial r} + \frac{1}{r^2}\frac{\partial^2}{\partial \varphi^2}\right), \quad \hat{H}_2 = -\frac{\partial^2}{\partial z^2}. \tag{4}$$

Here $\hat{H} = \frac{\hat{\mathrm{H}}}{E_R}$, $r = \frac{\rho}{a_B}$, $z = \frac{Z}{a_B}$, $E_R = \frac{\hbar^2}{2m_e^* a_B^2}$ is is the effective Rydberg energy of an electron, $a_B = \frac{\kappa\hbar^2}{m_e^* e^2}$ is the effective Bohr radius of an electron, e and m_e^* are the charge and effective mass of an electron, correspondingly, κ is a dielectric permittivity. The wave functions (WFs) of the problem is sought in the form

$$\Psi(r,\varphi,z) = Ce^{im\varphi}R(r;z)\chi(z), \tag{5}$$

where C is a normalization constant. At a fixed value of the coordinate z of the "slow" subsystem the particle motion is localized in a two-dimensional potential well with an effective variable width

$$r(z) = a\left(1 - \frac{|z|}{c}\right), \tag{6}$$

where $a = \frac{a_1}{a_B}$, and $c = \frac{c_1}{a_B}$. First, let us solve the Schrödinger equation of the "fast" subsystem, which after simple transformations reduces to the following Bessel equation:

$$R''(r;z) + \frac{1}{r}R'(r;z) + \left(\varepsilon(z) - \frac{m^2}{r^2}\right)R(r;z) = 0. \tag{7}$$

The solutions of the equation (7) are the first kind Bessel functions

$$R(r;z) = C\,J_m\left(\sqrt{\varepsilon(z)}\,r\right), \tag{8}$$

where C is a normalization constant, and m is a magnetic QN. From the boundary condition $J_m\left(\sqrt{\varepsilon(z)}\,r\right)\Big|_{r=r(z)} = 0$, and taking into account the expression (6), one obtains the energy of the "fast" subsystem:

$$\varepsilon(z) = \frac{\alpha_{n,m}^2}{a^2\left(1 - \frac{|z|}{c}\right)^2}, \quad n = 1,2,\ldots, \quad m = 0,1,\ldots. \tag{9}$$

For the lower levels of the energy spectrum, the CC motion is localized mainly in the vicinity of the geometrical center of the QD (interval $|z| \ll c$). Expanding $\varepsilon(z)$ in a power series around the point 0 these points one gets the expression

$$\varepsilon(z) \simeq \frac{\alpha_{n,m}^2}{a^2}\left(1 + \frac{2|z|}{c} + \frac{3z^2}{c^2}\right), \tag{10}$$

which serves as an effective potential for the "slow" subsystem. In contrast to the majority of cases of the adiabatic approximation applications (see e.g. [14,17]), in the expansion (10) the linear term is absent. Obviously, this is a consequence of the conical symmetry of the superellipsoidal QD. Therefore, it is important to consider in detail the impact of both linear and quadratic terms of the effective potential on the "slow" subsystem motion.

2.1.1 Linear approximation. The effective potential energy (10) of the "slow" subsystem in this case can be written as

$$\varepsilon(z) \simeq \frac{\alpha_{n,m}^2}{a^2}\left(1 + \frac{2|z|}{c}\right). \tag{11}$$

Since the linear term is the modulus of the coordinate (the triangular potential well), then the positive and negative z coordinates needs to be discussed separately. In case of positive coordinates $|z| = z$, the potential $\varepsilon(z) = \varepsilon_0 + fz$, where $\varepsilon_0 = \frac{\alpha_{n,m}^2}{a^2}, f = \frac{2\alpha_{n,m}^2}{a^2 c}$. The Schrödinger equation of the "slow" subsystem

$$\left(-\frac{\partial^2}{\partial z^2} + \varepsilon_0 + fz\right)\chi(z) = \varepsilon\chi(z), \tag{12}$$

where ε is the total energy of the electron in the QD. After change of the variable $\xi^+ = f^{\frac{1}{3}}z - f^{-\frac{2}{3}}\varepsilon_1$, one gets the Airy equation

$$\chi''(\xi) - \xi\chi(\xi) = 0, \tag{13}$$

where $\varepsilon_1 = \varepsilon - \varepsilon_0$. The solutions of (13) satisfying the standard conditions imposed on WFs at $z \to +\infty$, are Airy functions of the first kind:

$$\chi^+(z) = C_1 Ai\left(f^{\frac{1}{3}}z - f^{-\frac{2}{3}}\varepsilon_1\right). \tag{14}$$

In the case of negative coordinates $|z| = -z$, the potential has the form $\varepsilon(z) = \varepsilon_0 - fz$. Changing the variable $\xi^- = -f^{\frac{1}{3}}z - f^{-\frac{2}{3}}\varepsilon_1$, after the simple transformations the Schrödinger equation takes the form (13), and the solutions satisfying the standard conditions at $z \to -\infty$ can be written as

$$\chi^-(z) = C_2 Ai\left(-f^{\frac{1}{3}}z - f^{-\frac{2}{3}}\varepsilon_1\right), \tag{15}$$

where C_1 and C_2 are normalization constants. The electron total energy ε_{lin} in linear approximation can be determined from the condition of continuity of the WFs logarithmic derivatives:

$$\left.\frac{\chi'^+(z)}{\chi^+(z)}\right|_{z=0} = \left.\frac{\chi'^-(z)}{\chi^-(z)}\right|_{z=0.} \tag{16}$$

2.1.2 Quadratic approximation. Now consider the motion of the "slow" subsystem taking into account the parabolic term in the expansion (10). As in the previous case, consider the positive and negative z coordinates separately. In the case of positive z coordinates, the potential takes the form $\varepsilon(z) = \varepsilon_0 + fz + \gamma^2 z^2$, where $\gamma = \frac{\sqrt{3}\alpha_{n,m}}{ac}$. After certain transformations, the Schrödinger equation can be written as

$$\chi''(z) + \left(\varepsilon_2 - \gamma^2\left(z + \frac{f}{2\gamma^2}\right)^2\right)\chi(z) = 0, \tag{17}$$

where $\varepsilon_2 = \varepsilon - \varepsilon_0 + \frac{f^2}{4\gamma^2}$. Changing the variable $\eta^+ = \sqrt{2\gamma}\left(z + \frac{f}{2\gamma^2}\right)$ one gets the equation

$$\chi''(\eta) + \left(\frac{\varepsilon_2}{2\gamma} - \frac{\eta^2}{4}\right)\chi(\eta) = 0, \tag{18}$$

which solutions are the parabolic cylinder functions:

$$\chi^+(z) = C_1 D_\lambda\left(\sqrt{2\gamma}\left(z + \frac{f}{2\gamma^2}\right)\right), \tag{19}$$

where $\lambda = \frac{\varepsilon_2}{2\gamma} - \frac{1}{2}$. For the negative values of the coordinate $|z| = -z$, and the potential takes the form $\varepsilon(z) = \varepsilon_0 - fz + \gamma^2 z^2$. Then, changing the variable $\eta^- = \sqrt{2\gamma}\left(z - \frac{f}{2\gamma^2}\right)$, one gets the equation (18), which solutions are the parabolic cylinder functions:

$$\chi^-(z) = C_2 D_\lambda\left(\sqrt{2\gamma}\left(z - \frac{f}{2\gamma^2}\right)\right). \tag{20}$$

As in the case of the linear approximation, in the qudratic approximation one gets the total energy of the electron ε_{quad} from the condition of continuity of logarithmic derivatives (16) of the (19) and (20) WFs.

2.2 Weak SQ regime

Let us proceed to the weak SQ, when $a_{ex} \ll a_1$. Then the Coulomb interaction of the electron and the hole prevails over the SQ energy of the biconical QD walls in all geometrical directions. It means that the center-of-gravity motion of the exciton is quantized. The system's WFs can be represented as

$$\Psi(\vec{\rho}_e, \vec{\rho}_h) = \psi(\vec{\rho})\Phi_{n_r,l,m}(\vec{R}), \tag{21}$$

where $\vec{\rho} = \vec{\rho}_e - \vec{\rho}_h$, $\vec{R} = \frac{m_e^*\vec{\rho}_e + m_h^*\vec{\rho}_h}{m_e^* + m_h^*}$, m_h^* is the effective mass of the hole. The case of the heavy hole is considered $m_e^* \ll m_h^*$. In the expression (21) the $\psi(\vec{\rho})$ describes the relative motion of the electron and the hole, and $\Phi_{n_r,l,m}(\vec{R})$ describes the motion of the center-of-gravity of the exciton. Taking into account the Coulomb interaction, the system's Hamiltonian in $E_R^\mu = \frac{\hbar^2}{2\mu a_{ex}^2}$ units can be written as

$$\hat{H} = -\frac{\mu}{M}\nabla_{\vec{R}}^2 - \nabla_{\vec{r}}^2 - \frac{2}{|\vec{r}|}, \tag{22}$$

where $M = m_e^* + m_h^*$ is the mass of the exciton, $\mu = \frac{m_e^* m_h^*}{m_e^* + m_h^*}$ is the reduced mass of the exciton. After separation of variables and solving the Schrödinger equation for the energy spectrum and the WFs of the relative motion of the exciton one gets known results:

$$\varepsilon_{ex} = -\frac{\mu}{M}\frac{1}{S^2}, \quad S = 1, 2, \ldots, \tag{23}$$

$$\psi(r,\theta,\varphi) = r^l e^{-\frac{r}{2}} {}_1F_1(-S, 2l+2; r) Y_{l,m}(\theta,\varphi), \tag{24}$$

where S and l are the main and orbital QNs, correspondingly, ${}_1F_1(a,b;z)$ is the confluent hypergeometric function of the first kind, $Y_{l,m}(\theta,\varphi)$ are spherical functions [16]. One gets the energy ε_{Gr} of the motion of the exciton center of gravity, by repeating the above calculation procedure of the strong SQ regime for both linear and quadratic approximations, but with m_e^* being the exciton mass M. It should be noted that formed effective potentials (both linear and parabolic) does not affect the Coulomb interaction between the electron and the hole, and affect only to the motion of the center of gravity of the exciton. In other words, in the weak SQ regime, the quantization of the motion of an exciton in the QD does not affect the internal structure of the exciton. Finally, the total energy of the system in the weak SQ regime can be written as

$$\varepsilon = \varepsilon_{Gr} - \frac{\mu}{M}\frac{1}{S^2}, \tag{25}$$

where ε_{Gr} is the energy of the exciton center of gravity quantization, determined from the condition (16).

3. Direct interband absorption of light. Let us consider direct interband absorption of light in the QDM in the strong SQ regime, when the Coulomb interaction between the electron and a hole is neglected. The case of a heavy hole is discussed, when $m_e^* \ll m_h^*$. The absorption coefficient [19] is determined by the expression

$$K = A\sum_{\nu,\nu'}\left|\int \Psi_\nu^e \Psi_{\nu'}^h d\vec{r}\right|^2 \delta\left(\hbar\Omega - E_g - E_\nu^e - E_{\nu'}^h\right) \tag{26}$$

where ν and ν' are the QN sets corresponding to the electron and the heavy hole, E_g is the bandgap of a bulk semiconductor, Ω is an incident light frequency, A is a quantity proportional to the square of the matrix element taken by the Bloch functions. After simple calculations one gets the following expressions for the absorption edges (AE) of the linear and quadratic approximations W_{110} and W_{100}, correspondingly

$$W_{110} = 1 + \varepsilon_{lin}\frac{d^2}{a_B^2}, \quad W_{100} = 1 + \varepsilon_{quad}\frac{d^2}{a_B^2}, \tag{27}$$

where $W_{1\binom{0}{1}0} = \frac{\hbar\Omega_{1\binom{0}{1}0}}{E_g}$, and $d = \frac{\hbar}{\sqrt{2\mu E_g}}$. The formula (27) describes the dependence of the effective bandgap on the semiaxes a_1 and c_1. With decreasing both semiaxes, the AE shifts to the short wave region, however, the dependence on minor semiaxis is much stronger.

Consider now the selection rules in quantum transitions. Quantum transitions for the energy levels allowed for the magnetic QNs $m = -m'$, and $n = n'$ for the QNs of the "fast" subsystem. The condition of continuity of logarithmic derivatives of the WFs (16) leads to the complete removal of the "slow" subsystem QNs selection rules. Note, that the analytical expression (27) is given taking into account above mentioned selection rules. Next, consider the interband absorption of light in the weak SQ regime. Since the exciton is localized in a relatively small vicinity of the QD geometric center, the absorption coefficient can be written the following expression [19]:

$$K = A\sum_{n,n_r,l,m}\left|\varphi(0)\right|^2\left|\int \Phi_{n,n_r,m}\left(\vec{R}\right)d\vec{R}\right|^2 \delta\left(\hbar\Omega - E_g - E\right), \tag{28}$$

where E represents the energy (25) in dimensionless units. Note, that $\varphi(0) \neq 0$ only for the ground state, when $l = m = 0$ (l is the orbital QN of the exciton). In this regime one gets the following expressions for the AEs:

$$W_{1101} = 1 + \varepsilon_{Gr}^{lin}\frac{h^2}{a_B^2} - \frac{h^2}{a_{ex}^\mu a_{ex}^M}, \quad W_{1001} = 1 + \varepsilon_{Gr}^{quad}\frac{h^2}{a_B^2} - \frac{h^2}{a_{ex}^\mu a_{ex}^M}, \tag{29}$$

where $W_{\binom{0}{1}01} = \frac{\hbar\Omega_{\binom{0}{1}01}}{E_g}$, $h = \frac{\hbar}{\sqrt{2ME_g}}$, $a_{ex}^{\mu} = \frac{\kappa\hbar^2}{\mu e^2}$, and $a_{ex}^{M} = \frac{\kappa\hbar^2}{Me^2}$. The most important feature of the latter case is that the shift of the exciton level with changes in the semiaxes of the biconical QD is determined by the total mass of the exciton.

4. Discussion of results

As noted above, there is a nonzero linear term in the expansion (10) of the energy of the "fast" subsystem due to the particular symmetry of the biconical QD. In the absence of a linear term, the effective potential turns parabolic for most cases of applications of the geometrical adiabatic approximation [18]. This in turn leads to the appearance of the own equidistant families of sublevels over each energy level of the "fast" subsystem. However, in our case a non-zero linear term in the

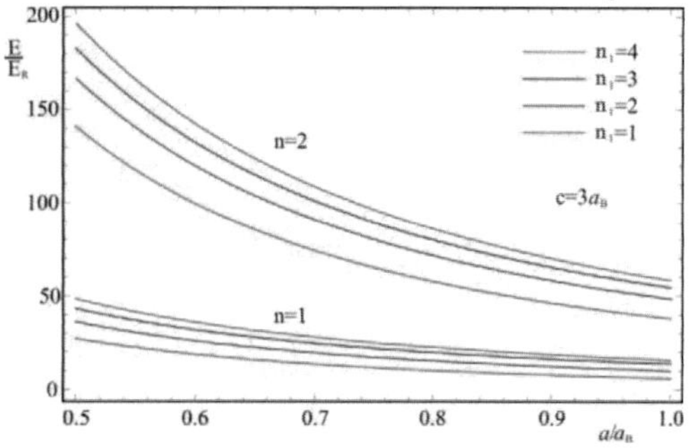

Figure 2. The dependence of the first two families of the electron energy levels on the minor semiaxis for the linear approximation.

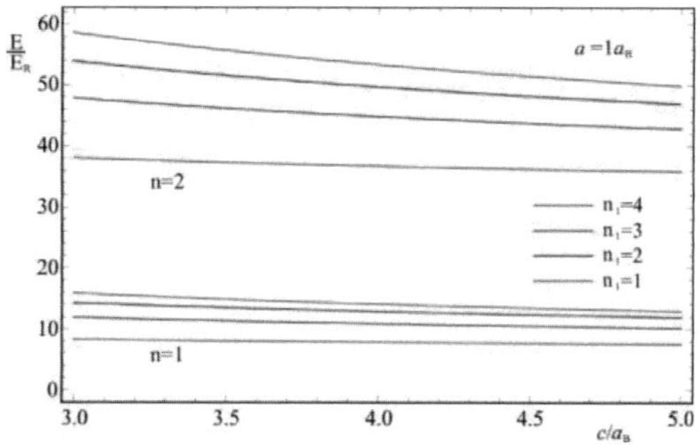

Figure 3. The dependence of the first two families of the electron energy levels on the major semiaxis for the linear approximation.

expansion (10) leads to the appearance of non-equidistant sublevels. Figure 2 represents the dependence of the first two families of the energy levels of the CC in the biconical QD on the minor semiaxis a at a given value of the major semiaxis c in the case of the linear approximation. As it is seen from the figure, the sublevels of the CC are non-equidistant. It should be noted that in this case also, every level of the "fast" subsystem has its own family of sublevels of the "slow" subsystem, with interlevel distances depending on the "fast" subsystem n QN. With an increase in the semiaxis the energy levels decrease. At the same time, interlevel distances of the energy families decrease, as a consequence of the QD walls' SQ reducing in radial direction.

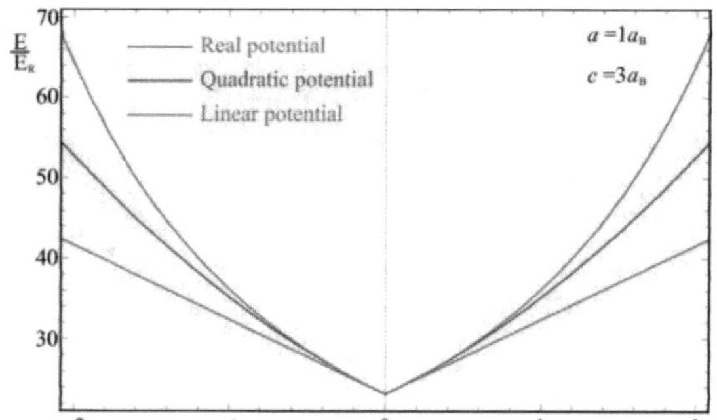

Figure 4. The dependences of linear and quadratic effective potentials and the real potential on the "z" coordinate in the biconical QD.

A similar dependence of the electron energy on the major semiaxis c of the biconical QD is shown in the figure 3. As the figure shows, the increase in the major semiaxis leads to a significantly weak dependence due to small contribution of the SQ effects, in contrast to the dependence on the minor semiaxis. In other words, the SQ effects appear brighter in the radial direction as a consequence of the condition $a_1 \ll c_1$, as expected. On the other hand, the effect of the linear term in the effective potential on the CC motion is stronger only for the first two families of the lower energy levels of the spectrum. Figure 4 shows the dependence of the effective and the real potentials on the coordinate z. As it is seen from the figure, the linear part perfectly is converged and approximated with the real potential only at small values of the coordinate z. For higher values of the energy, one must take into account the parabolic term, since the linear term strongly diverges the curve of the real effective potential. Obviously, accounting the parabolic term leads to more successful approximation of the real potential. However, it leads to reconstruction of the lower levels of the CC spectrum also. The presence of the parabolic term in (10) leads to a narrowing of the CC localization area and an enhancement of the SQ influence along the OZ axis.

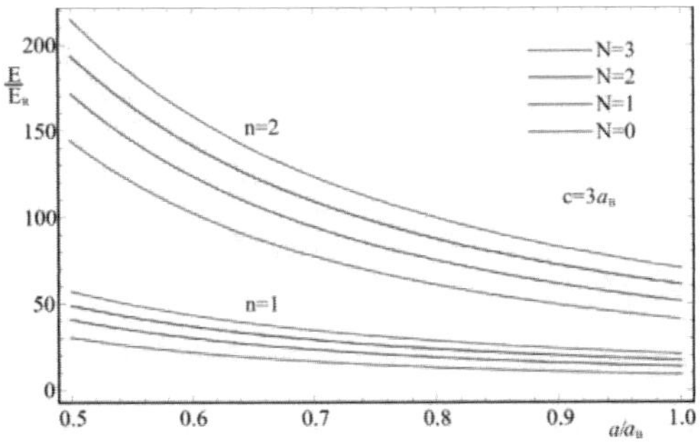

Figure 5. The dependence of the first two families of the electron energy on the minor semiaxis for the linear approximation.

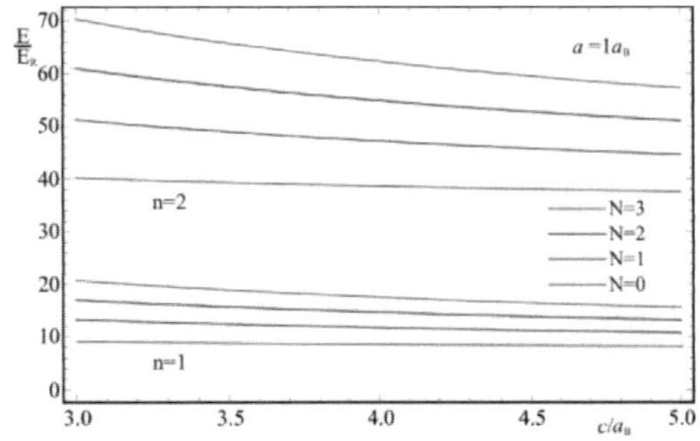

Figure 6. The dependence of the first two families of the electron energy on the major semiaxis for the quadratic approximation.

Figure 5 plots the dependence of the first two families of the CC energy levels in the biconical QD on the minor semiaxis a at the given value of the major semiaxis for the linear approximation. As noted above, the equidistant sublevels of the "slow" subsystem are positioned over each level of the "fast" subsystem. It can be seen that interlevel distances of families increases with transition from the first to the second family, as expected, due to the interlevel distances dependence on the "fast" subsystem QN n. Interlevel distances also depend on the value of the semiaxes of biconical QD. With increasing the minor semiaxis the distance between the curves decreases, due to a decrease in the SQ impact in the radial direction. In this case also, the dependence of the CC energy on the small semiaxis appears brighter. However, the interlevel distances of equidistant levels do not depend on the mass of the CC, in contrast to the case of the quantum oscillator. As a consequence, the area of the CC localization is also does not dependent on the mass of the particle. Thus, the characteristic length of a quantum oscillator is $a_\omega = \sqrt{\frac{\hbar}{m_e \omega}}$, where ω is the frequency of an oscillator. In our case, the parabolic potential arises due to the application of the adiabatic approximation, and the localization area is $a_\omega^{ad} = \sqrt{\frac{\hbar}{m_e^* \omega_{ad}}} = \sqrt{\frac{\sqrt{3}\alpha_{n,m}}{ac}}$; particle localization condition in the QD geometric center does not on the mass of the particle, but only on the geometrical parameters of the QD.

Weaker expressed energy dependence of the first two families of the equidistant spectrum on the major semiaxis of the biconical QD is shown on the figure 6. As it is seen, with an increase in the semiaxis c, the energy curves decrease slowly, due to the SQ weak influence in the perpendicular direction than in the radial direction. Thus, at $a = 1a_B$ and $c = 3.5a_B$, the interlevel distance for the first energy family $(n=1)$ is $\Delta E \simeq 3.78E_R$, whereas at $c = 5a_B$ one has $\Delta E \simeq 3.17E_R$. For comparison, at $c = 3a_B$ and $a = 0.5a_B$ (see figure 5) one has $\Delta E \simeq 10.43E_R$ (the transition frequency is $\omega = 8.352 \cdot 10^{13}\ \text{sec}^{-1}$ and corresponds to the infrared region of the spectrum), but at $a = 0.8a_B$ one has $\Delta E \simeq 6.75E_R$ (the transition frequency is $\omega = 5.405 \cdot 10^{13}\ \text{sec}^{-1}$). All numerical calculations are done for the *GaAs* QD with the following parameters: $m_e^* = 0.067 m_e$, m_e is a free electron mass, $\kappa = 13.8$, $E_R = 5.275 meV$ and $a_B = 104 A^\circ$.

An evident illustration of the difference in the linear and quadratic cases is demonstrated on the figure 7, which plots the dependence of the first two levels of the total energy of the electron on the minor semiaxis a at the given values of the major semiaxis c. With increase in the minor semiaxis the curves decrease in both cases of the effective potential approximation, and the difference between them reduces. This is due to the SQ impact reduction in radial direction. The curves

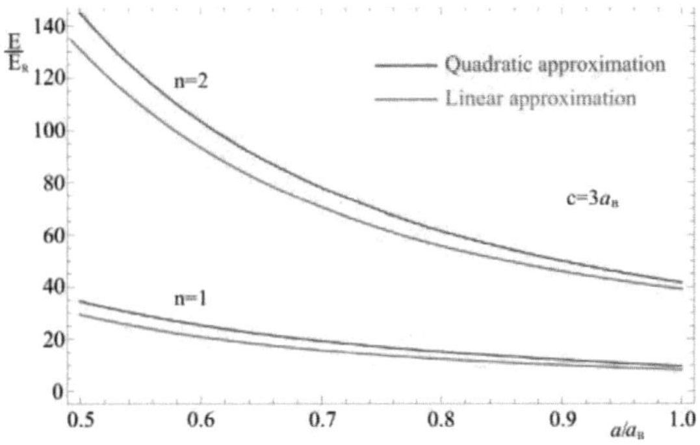

Figure 7. The dependence of the first two levels of the total energy of the electron on the minor semiaxis for the linear and quadratic approximations of the effective potential.

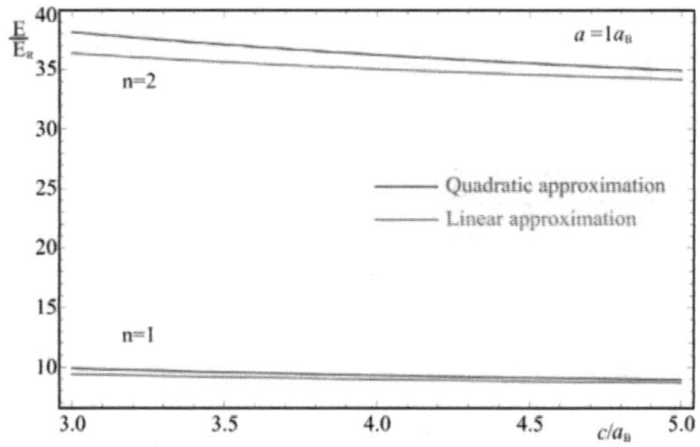

Figure 8. The dependence of the first two levels of the total energy of the electron on the major semiaxis for the linear and quadratic approximations of the effective potential.

corresponding to the parabolic approximation are positioned higher, as expected, since the CC localization area is smaller than in the triangular quantum well, corresponding to the linear approximation (see figure 4). For small values of the semiaxis, the SQ effect is stronger, and the particle experiences the difference between the triangular and parabolic wells more stronger, which results in curves discrepancy. For large values of the a, this effect weakens, and the curves

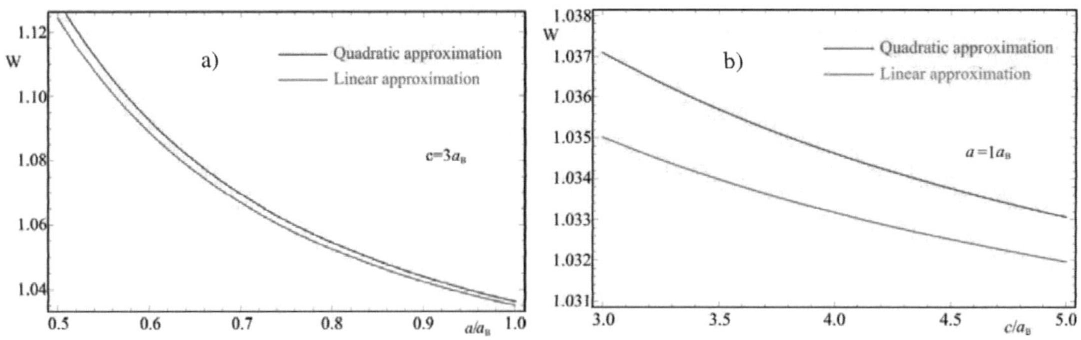

Figure 9. The dependences of the AE for the linear and quadratic approximations (a) on the minor semiaxis, (b) on the major semiaxis.

corresponding to both approximations merges. A similar pattern is observed in the figure 8, which shows the dependence of the first two levels of the total energy of the electron on the value of the major semiaxis c at the given a. Again, the energy curves of the CC corresponding to the parabolic approximation are positioned higher due to stronger SQ impact of the QD wells.

Figure 9 shows the dependences of the AE of the of direct interband absorption of light on the biconical QD minor and major semiaxes, respectively, in the strong SQ regime. With a decrease in semiaxes, the AE increases, which is a consequence of the SQ increase (the "effective" banbgap increases). As can be seen from the figures, the AE change appears brighter depending on the QD minor semiaxis. In both figures the curve of the parabolic case AE is positioned higher (blueshift) than the AE curve of linear approximation. This blueshift of the edge frequencies is the result of an

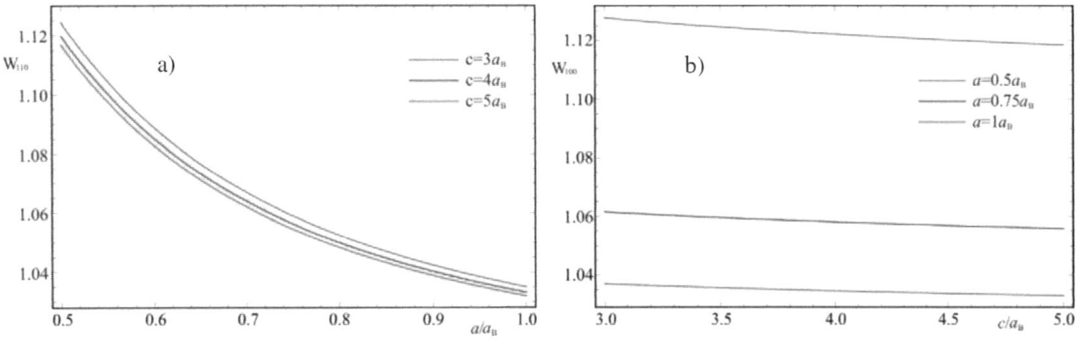

Figure 10. The dependences of the AE (a) on the minor semiaxis at the different "c", (b) on the major semiaxes at the different "a".

increase in the effective bandgap due to the stronger confinement in the case of a parabolic approximation. Similar blueshift is observed in the case of a parabolic approximation for weak SQ regime, however, the confinement is not prevalent. In the weak SQ regime the exciton forms, which motion as a whole is influenced by the weak confinement of the QD walls. The accounting of Coulomb interaction leads to a decrease in the "effective" bandgap and a redshift of the AE (the negative excitonic term in (29)). Coulomb interaction between the electron and a hole results in the appearance of energy levels in the bandgap, which leads to the AE shift to the long wavelength region.

Finally, figure 10 illustrates the dependences of the AE on the biconical QD minor and major semiaxes in strong SQ regime at the given values of the major and minor semiaxes, correspondingly. As one can see, even a large variation in the semiaxis c leads to a small change in the edge frequencies, whereas even a small variation in the minor semiaxis a results in a sharp leap in the AE value.

5. Conclusion

In the framework of the geometrical adiabatic approximation the electronic states in the *GaAs* biconical QD have been theoretically investigated in a strong and weak SQ regimes. An atypical linear term in the "slow" subsystem effective confining potential has been revealed as a consequence of the QD's conical symmetry. It has been shown that the linear term in the effective potential forms a family of non-equidistant sublevels on each level of the "fast" subsystem in the electron energy spectrum, whereas the quadratic term leads to the spectrum reconstruction and appearance of the equidistant energy levels. The localization area or the characteristic length of oscillator has been shown to be independent on the particle mass (electron, hole, exciton). The dependences of the AE on the QD semiaxes values and the order of approximation of the effective potential have been studied as well. The appearance of additional Coulomb sublevels in the QD has been shown to result in the edge frequencies redshift, while the quadratic approximation leads to an additional AE blueshift. Partial withdrawal of the selection rules in quantum transitions has been revealed as well.

Acknowledgements
This work is supported by the NSF (HRD-1345219), NASA (NNX09AV07A) and in part by the M. Hildred Blewett Fellowship of the American Physical Society, www.aps.org.

References

[1] Harrison P 2005 *Quantum Wells, Wires and Dots, Theoretical and Computational Physics* (New York: Wiley)

[2] Bimberg D, Grundmann M, and Ledentsov N 1999 *Quantum Dot Heterostructures* (New York: Wiley)

[3] Dvoyan K G and Kazaryan E M 2001 Impurity states in a weakly prolate (oblate) ellipsoidal microcrystal placed in a magnetic field *Phys Status Solidi B* **228** 695-703

[4] Sadeghi E and Naghdi E 2015 Polarization charges effects on optical properties of double ellipsoidal quantum dot *Physica B* **457** 178-181

[5] Dvoyan K G, Kazaryan E M, and Petrosyan L S 2005 Electronic states in quantum dots with ellipsoidal symmetry *Physica E* **28** 333-338

[6] Gambaryan K M, Aroutiounian V M, Boeck T, Schulze M, and Soukiassian P 2008 Strain-induced InAsSbP islands and quantum dots grown by liquid phase epitaxy on an InAs(100) substrate *J Phys D: Appl Phys* **41** 162004-162009

[7] Shi L and Yan Z 2013 Effects of electric field and shape on the ground state energy of an exciton bound to an ionized donor impurity in ellipsoidal quantum dots *J Appl Phys* **114** 194301

[8] Aroutiounian V M, Gambaryan K M, Harutyunyan V G, Soukiassian P G, Boeck T, Schmidtbauer J, and Bansen R 2013 The Ostwald ripening at nanoengineering of InAsSbP spherical and ellipsoidal quantum dots on InAs (100) surface *J Contemp Phys Arme+* **48** 37-42

[9] Sadeghi E and Alirezaie S 2013 Effect of incident light polarization on optical properties of an ellipsoidal quantum dot *Superlattices Microstruct* **54** 128-136

[10] Dvoyan K G, Evoyan V V, Kazaryan E M, Nazmitdinov R G, and Sarkisyan H A 2012 Magneto-absorption in ellipsoidal quantum dot *Int J Mod Phys Conf Ser* **15** 40-47

[11] Sablon K 2009 Ellipsoidal InAs quantum dots *Nanoscale Res Lett* **4** 1256

[12] Cantele G, Ninno D, and Iadonisi D 2000 Confined states in ellipsoidal quantum dots *J Phys: Condens Matter* **12** 9019-9036

[13] Landau L D and Lifshitz E M *Quantum Mechanics* 1989 (Moscow: Nauka)

[14] Dvoyan K G, Hayrapetyan D B, Kazaryan E M, and Tshantshapanyan A A 2007 Electron states and light absorption in strongly oblate and strongly prolate ellipsoidal quantum dots in presence of electrical and magnetic fields 2007 *Nanoscale Res Lett* **2** 601-608

[15] Hayrapetyan D B, Dvoyan K G, and Kazaryan E M 2006 Light absorption in ellipsoidal quantum dots, *PEM Divnomorskoe, Russia, September 24-29*: Proc 10-*th international conference* **2** 74-76

[16] Dvoyan K G, Kazaryan E M, and Tshantshapanyan A A 2009 Light absorption in coated ellipsoidal quantum lenses J Mater Sci - Mater Electron **20** 491-498

[17] Dvoyan K G, Kazaryan E M, and Tshantshapanyan A A 2007 Electronic states and light absorption in thin ellipsoidal quantum lens under the influence of strong magnetic field *J Contemp Phys Arme+* **42** 282-286

[18] Kazaryan E M, and Dvoyan K G 2010 Ellipsoidal and lens-shaped quantum dots (Nanoscience and Nanotechnology. Encyclopedia of Life Support Systems. UNESCO, EOLSS), pp 134-149

[19] Efros Al L and Efros A L 1982 *Sov Phys Semicond* **16** 772

Conventional and Unconventional Pairing and Condensates in Dilute Nuclear Matter

John W Clark[1,2], **Armen Sedrakian**[3], **Martin Stein**[3], **Xu-Guang Huang**[4], **Victor A Khodel**[1,5], **Vasily R Shaginyan**[6], **Mikhail V Zverev**[5,7]

[1] Department of Physics and McDonnell Center for the Space Sciences, Washington University, St. Louis, MO 63130 USA
[2] Center for Mathematical Sciences, University of Madeira, Funchal, 9000-390 Portugal
[3] Institute for Theoretical Physics, J. W. Goethe-University, D-60438 Frankfurt am Main, Germany
[4] Physics Department & Center for Particle Physics and Field Theory, Fudan University, Shanghai 200433, China
[5]National Research Centre Kurchatov Institute, Moscow, 123182, Russia
[6] Petersburg Nuclear Physics Institute, NRC Kurchatov Institute, Gatchina, 188300, Russia
[7] Moscow Institute of Physics and Technology, Dolgoprudny, Moscow District 141700, Russia

E-mail: jwc@wuphys.wustl.edu

Abstract. This contribution will survey recent progress toward an understanding of diverse pairing phenomena in dilute nuclear matter at small and moderate isospin asymmetry, with results of potential relevance to supernova envelopes and proto-neutron stars. Application of *ab initio* many-body techniques has revealed a rich array of temperature-density phase diagrams, indexed by isospin asymmetry, which feature both conventional and unconventional superfluid phases. At low density there exist a homogeneous translationally invariant BCS phase, a homogeneous LOFF phase violating translational invariance, and an inhomogeneous translationally invariant phase-separated BCS phase. The transition from the BCS to the BEC phases is characterized in terms of the evolution, from weak to strong coupling, of the pairing gap, condensate wave function, and quasiparticle occupation numbers and spectra. Additionally, a schematic formal analysis of pairing in neutron matter at low to moderate densities is presented that establishes conditions for the emergence of both conventional and unconventional pairing solutions and encompasses the possibility of dineutron formation.

1. Introduction

This report serves to review and analyze a body of recent findings on the phase diagram of dilute nuclear matter, calculated over wide ranges of density, temperature, and isospin asymmetry. Quantitative results will be presented for the temperature-density $(T-\rho)$ phase diagram at baryon densities below about half the saturation density of isospin-symmetric nuclear matter, but pairing phenomena that may occur at somewhat higher densities will also be addressed. The corresponding studies [1, 2, 3, 4, 5], both theoretical and numerical, focus attention on the emergence of unconventional as well as conventional pairing in the $^3S_1$-$^3D_1$ (deuteron) channel as well as associated BCS-BEC crossovers. Earlier work on these and closely related themes has been described in [6, 7, 8, 9, 10, 11] and more recently in [12, 13, 14, 15, 17, 18, 19].

Journal of Physics: Conference Series **702** (2016) 012012 doi:10.1088/1742-6596/702/1/012012

Application of *ab initio* quantum many-body theory to this problem domain has the distinct advantage that, within the density regime considered in the numerical study, the two-body nucleon-nucleon (NN) interactions are well constrained by the NN phase-shift data and the properties of the deuteron. Nor is the problem purely academic, as it is directly relevant to the matter existing in supernovae envelopes and proto-neutron stars (having relatively low temperatures and low isospin asymmetries) and in neutron star crusts (cold, with large isospin asymmetries).

The complex phenomenology of dilute nuclear matter, summarized in its $T-\rho$ phase diagram determined over a range of isospin asymmetries, arises from three sources:

(1) Pauli exclusion acting for fermionic species (nucleons, tritons, ^{3}He, etc.) and Bose-Einstein condensation (BEC) of bosonic species (such as deuterons and alpha particles). These are most effective at high particle densities (but below nuclear saturation), low temperatures, and low mass number of nuclear species.

(2) Dominance of the longer-range attractive component of the NN interaction. At low densities and not-so-low temperatures, this component is responsible for the formation of tightly bound nuclear clusters (deuterons, dineutrons(?), tritons, alphas, ...), which can undergo BEC in the case of bosonic clusters. At higher densities and low temperatures it is responsible for the formation of Cooper pairs with pairing gap Δ. At small isospin asymmetries the pairing is in the triplet $^3S_1$-$^3D_1$ channel, whereas for large asymmetries the pairing is in the $^1S_0$ singlet channel.

(3) Isospin asymmetry, induced by weak interactions, producing a mismatch of neutron (n) and proton (p) Fermi momenta, giving rise to mixed superfluid/normal phases and unconventional pairing – Cooper pairs with nonzero center-of-mass (CM) momentum (the so-called LOFF phase [20]) or deformed neutron and proton Fermi surfaces [21].

2. Phase Diagram of Dilute Isospin-Symmetric Nuclear Matter

We begin with a brief examination of the low-density phase diagram for the fiducial case of isospin-symmetric nuclear matter with equal neutron and proton baryon densities, $\rho_n = \rho_p$, as explored in [1]. A model of low-density nuclear matter based on a simplified two-nucleon interaction was considered which exhibits behavior generic to systems of fermions interacting via a short-range repulsion and a longer-range attraction. Such behavior includes both (i) formation of clusters tightly bound in the medium at lower densities and higher temperatures, and (ii) Cooper pairing in a BCS state at lower temperatures and higher densities. Specifically, a Malfliet-Tjon model with MF-III parametrization [22] was chosen for the NN interaction, consisting of a central but spin-dependent superposition of inner repulsive and outer attractive Yukawas, fitted to NN S-wave phase shifts and deuteron binding. This interaction shows a strong pairing instability in the $^3S_1$ channel.

To study the low-temperature superfluid phase, the BCS gap equation was solved self-consistently for the energy gap $\Delta(T)$ and the chemical potential μ below the critical temperature T_{sc}, with results shown in Fig. 1. Proceeding from higher to lower densities in the domain under study, conditions range from weak coupling (WC) to strong coupling (SC) as measured by the ratio $\Delta(0)/|\mu|$; a change of sign of the chemical potential from positive to negative is a signature of the WC $\rightarrow$ SC transition. In the low-ρ limit the gap equation reduces to the Schrödinger equation for the two-body bound state, with energy eigenvalue given by 2μ, which is naturally identified with the deuteron and subject to Bose-Einstein condensation. Accordingly, this earlier study provides a model of the BCS-BEC transition from BCS Cooper pairing in the $^3S_1$ state to a Bose condensate of simplified deuterons.

To extend the phase diagram to higher temperatures, the Lippmann-Schwinger and Faddeev equations were adapted to solve two- and three-nucleon bound-state problems in the presence

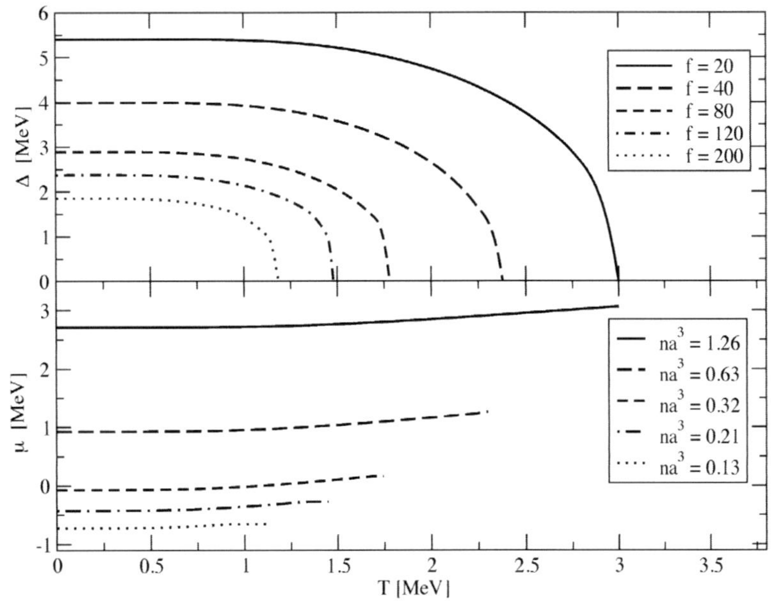

Figure 1. Dependence of the pairing gap (upper panel) and chemical potential (lower panel) on temperature for fixed values of $f = \rho_0/\rho$, where $n \equiv \rho$ denotes the baryon density and $\rho_0 = 0.16$ fm^{-3} the saturation density of symmetrical nuclear matter. Values of the dimensionless density parameter $n|a|^3$ assume a scattering length $a = 5.4$ fm.

of a dispersive fermionic background medium and attendant Pauli blocking effects. Evolution of clustering into deuteron dimers and triton and helion trimers was followed under increasing temperature and/or decreasing density. For small temperatures the quantum degeneracy is large and Pauli blocking strongly suppresses the binding energy of these clusters, which are quenched at a common critical temperature T_{cc}.

The results on pairing and bound states are combined to produce the schematic phase diagram in Fig. 2, showing several distinct regions in the $T - \rho$ plane:

G. The gaseous region above the solid critical line $T_{cc}(\rho)$ is populated by trimers, along with np dimers at lower temperatures.

B. The low-temperature, low-density domain (lower left: $\rho|a|^3 \ll 1$) contains a Bose condensate of tightly-bound deuterons.

C. The low-temperature, high-density domain (lower right: $\rho|a|^3 \gg 1$) features a BCS condensate of weakly-bound Cooper pairs.

L. The domain between the two critical lines $T_{cc}(\rho)$ and $T_{sc}(\rho)$ contains normal nucleonic $\{p, n\}$ liquid.

The superfluid phases labeled B and C are characterized by broken symmetry associated with the $\langle\psi\psi\rangle$ condensate. The transition C to B does not involve symmetry changes – it is a smooth crossover from the BCS to the BEC condensate. B $\to$ L and C $\to$ L are second-order phase transitions related to the vanishing of the condensate along the line $T_{sc}(\rho)$. The transition G $\to$ L (e.g., at vertical line) is characterized by an order parameter given by the fraction of trimers, which goes to zero at $T_{cc}(\rho)$ (tentatively second order).

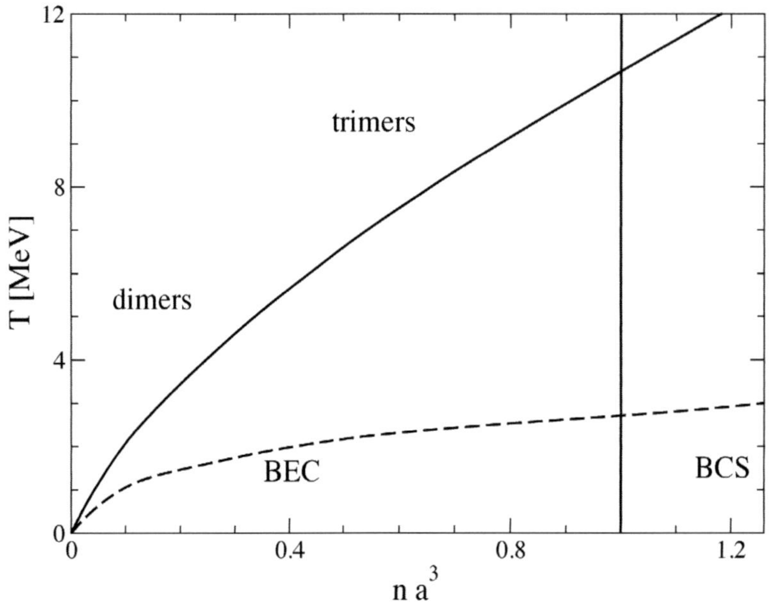

Figure 2. Phase diagram of dilute isospin-symmetric nuclear matter. Solid line: Critical temperature for extinction of three-body bound states; trimers and dimers exist above this line. Dashed line: Critical temperature for destruction of condensate. Weakly coupled BCS superfluid exists below and far to right of vertical line, Bose-Einstein condensate (BEC) of tightly bound np pairs exists below and far to the left. Dimensionless density $n|a|^3$ (with $n \equiv \rho$) as defined in Fig. 1.

3. Effects of Isospin Asymmetry on the Phase Diagram

The phase diagram of dilute nuclear matter becomes much more complex upon introduction of isospin asymmetry as another control variable, measured by

$$\alpha = (\rho_n - \rho_p)/(\rho_n + \rho_p) \tag{1}$$

in terms of neutron and proton number densities ρ_n and ρ_p. Unconventional superfluid and heterogeneous phases then emerge, largely dictated by mismatch of neutron and proton Fermi momenta k_{Fn} and k_{Fp}, which entails incomplete overlap of spherical neutron and proton Fermi spheres. As is well known, this mismatch may be mitigated by deformation of the n and p Fermi spheres so as to increase their phase-space overlap [21]. Otherwise, phase-space overlap may be enhanced by the formation of Cooper pairs having non-zero CM momentum [20, 21]. This is the alternative explored quite thoroughly in the recent work that is the primary subject of this paper. BCS pairing theory will be generalized to this case in the next section. Increase of temperature is another option for compensating the mismatch of k_{Fn} and k_{Fp}, due to the smearing of both Fermi surfaces. This effect, as well as the energetic advantage of dimerization at lower densities, gives rise to heterogeneous phases with superfluid and normal components.

The problem now has two energy scales: the pairing gap Δ in the ${}^3S_1$-${}^3D_1$ channel, realistic NN interactions now being employed, and the shift $\pm\delta\mu = \pm(\mu_n - \mu_p)/2$ between chemical potentials of neutrons and protons. With increasing isospin asymmetry, $|\delta\mu|$ increases from zero to values of order Δ, and a sequence of unconventional phases will appear.

4. Gap Equation for Nonzero CM Momentum

A BCS suitable gap equation allowing for Cooper pairs with nonzero CM momentum may be derived in quasiparticle approximation in the framework of imaginary-time finite-temperature Green's functions and the Nambu-Gor'kov basis [2, 4, 3]. The resulting quasiparticle spectra are written in a general reference frame moving with CM momentum $\mathbf{Q}$ relative to a laboratory rest frame.

This *ab initio* many-body theory yields solutions of the form

$$G^{\pm}_{n/p} = \frac{ik_\nu \pm \epsilon^{\mp}_{p/n}}{(ik_\nu - E^{+}_{\mp/\pm})(ik_\nu + E^{-}_{\pm/\mp})}, \tag{2}$$

$$F^{\pm}_{np} = \frac{-i\Delta}{(ik_\nu - E^{+}_{\pm})(ik_\nu + E^{-}_{\mp})}, \qquad F^{\pm}_{pn} = \frac{i\Delta}{(ik_\nu - E^{+}_{\mp})(ik_\nu + E^{-}_{\pm})}, \tag{3}$$

for the normal and anomalous Green's functions, respectively, where $k = (ik_\nu, \mathbf{k})$, with $k_\nu = (2\nu + 1)\pi T$, ν being any integer. There are four quasiparticle spectral branches specified by

$$\epsilon^{\pm}_{n\uparrow/\downarrow} = E_S - \delta\mu \pm E_A, \qquad \epsilon^{\pm}_{p\uparrow/\downarrow} = E_S + \delta\mu \pm E_A, \tag{4}$$

where, with $a, r \in \{+,-\}$,

$$E^{a}_{r} = \sqrt{E_S^2 + \Delta^2} + r\delta\mu + aE_A, \qquad E_S = (Q^2/4 + k^2)/2m^* - \bar{\mu}, \qquad E_A = \mathbf{k}\cdot\mathbf{Q}/2m^*. \tag{5}$$

The isospin asymmetry enters through the parameter $\delta\mu = (\mu_n - \mu_p)/2$, while $\bar{\mu}$ is the mean chemical potential.

In mean-field approximation, the anomalous self-energy (pairing gap) is expressed as

$$\Delta(\mathbf{Q}) = \frac{1}{4\beta}\int \frac{d^3k'}{(2\pi)^3}\sum_\nu V(\mathbf{k},\mathbf{k}')\mathrm{Im}[F^{+}_{np}(k'_\nu,\mathbf{k}',\mathbf{Q}) + F^{-}_{np}(k'_\nu,\mathbf{k}',\mathbf{Q}) - F^{+}_{pn}(k'_\nu,\mathbf{k}',\mathbf{Q}) - F^{-}_{pn}(k'_\nu,\mathbf{k}',\mathbf{Q})] \tag{6}$$

in terms of the above anomalous propagators, where $V(\mathbf{k},\mathbf{k}')$ is the np pairing interaction. The next steps toward deriving the gap equation as used in the numerical study of pairing in dilute nuclear matter involve (i) evaluation of the Matsubara sum over ν in the above expression, (ii) performing a partial wave expansion, and (iii) restricting attention to the $^3S_1$-$^3D_1$ channel, in which the appropriate np interaction is denoted by $V_{l,l'}(k,k')$ with $l, l' = 0, 2$. Thus one arrives at

$$\Delta_l(Q) = \frac{1}{4}\sum_{a,r,l'}\int \frac{d^3k'}{(2\pi)^3}\frac{V_{l,l'}(k,k')\Delta_{l'}(k',Q)}{2\sqrt{E_S^2(k') + \Delta^2(k',Q)}}[1 - 2f(E^{r}_{a})], \tag{7}$$

wherein $f(E^r_a) = 1/[\exp(E^r_a/T) + 1]$ and $\Delta^2 = (3/8\pi)\sum_l \Delta_l^2$.

Correspondingly, the partial number densities $\rho_{n/p}(\mathbf{Q})$ of neutrons and protons are determined in terms of normal propagators by

$$\rho_{n/p}(\mathbf{Q}) = \frac{2}{\beta}\int \frac{d^3k}{(2\pi)^3}\sum_\nu G^{+}_{n/p}(k_\nu,\mathbf{k},\mathbf{Q}) = 2\int \frac{d^3k}{(2\pi)^3}\frac{1}{2}\left[(1+\xi)f(E^{+}_{\mp}) + (1-\xi)f(-E^{-}_{\pm})\right], \tag{8}$$

where $\xi = E_S/\sqrt{E_S^2 + \Delta^2}$.

5. Calculational Specifics and Interactions Assumed

The coupled gap equations and the two density equations were solved self-consistently for a bare pairing interaction in the $^3S_1$-$^3D_1$ partial wave, as provided by a phase-equivalent in-vacuum NN interaction, namely by the Paris potential, thus implying Cooper pairing in the $S = 1$, $T = 0$ spin-isospin channel. As needed, the nuclear mean field is modeled by a Skyrme density functional, with SkIII and SLy4 parametrizations yielding nearly identical results.

Two simplifications are made:

(i) Polarization effects, i.e., medium modification of the input NN interaction (due for example to virtual exchange of density and spin-density excitations) are neglected, although they are known to be important in some regions of the phase diagram.

(ii) Apart from deuteron dimerization in the BEC phases, effects of nuclear clustering are not considered, although at somewhat higher temperatures one expects substantial populations of tritons, ^{3}He nuclei, and α particles, along with deuterons.

It should also be noted that $^1S_0$ Cooper pairing in the $S = 0$, $T = 1$ spin-isospin channel may mix and eventually replace $^3S_1$-$^3D_1$ pairing at asymptotically low T (below 0.5 MeV) and high asymmetry.

6. Free-Energy Minimization

The phase at each point in the $T - \rho$ phase diagram is determined by minimization of the free energy. In the case of pre-BEC homogeneous phases (perhaps translationally noninvariant), there are three possibilities: (i) $Q = 0$, $\Delta \neq 0$ (BCS phase), (ii) $Q \neq 0$, $\Delta \neq 0$ (LOFF phase), and (iii) $Q = 0$, $\Delta = 0$ (unpaired, normal phase). The ground state is determined by minimization of the free energy $F = E - TS$ of the superfluid ($\mathcal{S}$) or unpaired normal phase ($\mathcal{N}$) with respect to the parameter Q, where E is the internal energy determined from the Hamiltonian and S the entropy. Stability of the superfluid phase requires $F_\mathcal{S} < F_\mathcal{N}$.

As already indicated, there can also be a (pre-BEC) heterogeneous, phase-separated phase, denoted PS-BCS. Its free energy takes the form of a linear combination of superfluid and unpaired free energies,

$$\mathcal{F}(x, \alpha) = (1 - x)F_S(\alpha = 0) + xF_N(\alpha \neq 0),$$

which is to be minimized with respect to the filling fraction x of the unpaired component. The net densities of n/p per unit volume are given by $\rho_{n/p} = (1 - x)\rho^{(S)}_{n/p} + x\rho^{(N)}_{n/p}$. In the pure $\mathcal{S}$ phase, $\rho^{(S)}_n = \rho^{(S)}_p = \rho^{(S)}/2$.

7. Overview of the Phase Diagram

In this section we present an overview of the diverse phases that arise as the calculation proceeds from higher to lower densities in dilute nuclear matter at chosen isospin asymmetries $\alpha \geq 0$.

7.1. Conventional Phases

As specified, the microscopic calculation yields a smooth crossover from the pure BCS phase to an asymptotic state corresponding to a mixture of a deuteron BEC and a normal gas of the left-over unpaired neutrons. The transition from BCS to BEC is identified by two criteria: (i) The average chemical potential $\bar{\mu}$ changes its sign from positive to negative values, and (ii) The coherence length ξ of a Cooper pair becomes comparable to the interparticle distance d,

$$\xi \sim d = (3/4\pi\rho)^{1/3},$$

as ξ ranges from $\xi \gg d$ to $\xi \ll d$.

Journal of Physics: Conference Series **702** (2016) 012012 doi:10.1088/1742-6596/702/1/012012

7.2. Unconventional Phases: LOFF and PS-BCS

At $\alpha \neq 0$ the LOFF state can emerge due to the energetic advantage gained with a condensate that breaks translational symmetry: Cooper pairs that carry a nonzero CM momentum $\mathbf{Q}$ can compensate for the mismatch of neutron and proton chemical potentials. The calculations reveal the existence of a LOFF-state gap of isospin-asymmetric nuclear matter in a narrow regime at relatively low T and relatively high ρ values, having a maximum at finite Q, implying maximum condensation energy for such pairs. At large α, the maximum gap occurs at large values of Q. On the other hand, at a given asymmetry, an increase of temperature shifts the gap maximum and free-energy minimum toward smaller Q. With reduction of asymmetry, increase of temperature, and/or decrease of density, the BCS phase regains favor over the LOFF phase, with PS-BCS phase having the advantage in the contest at lower temperatures.

This behavior relative to the LOFF phase is well understood in terms of the phase-space overlap of the Fermi surfaces of neutrons and protons, which increases with increasing temperature on the one hand, and with momentum Q on the other. Similarly, the PS-BCS phase, in which a standard ($^3S_1$-$^3D_1$) BCS component coexists with a normal Fermi liquid made up of the excess unpaired neutrons, is clearly favored energetically relative to pure BCS and LOFF phases at the lowest temperatures, where there is little benefit from eroded Fermi surfaces of from forfeit of translational invariance.

What remains is an account of how the BCS-BEC crossover is affected by the existence of the unconventional nuclear LOFF and PS phases at nonzero α, under decrease of system density.

8. Inventory of Condensed (and Uncondensed) Phases

With the density conveniently measured as $\log(\rho/\rho_0)$, results from the numerical calculations for different asymmetries α are plotted in the composite phase diagram of Fig. 3, which shows several distinct phases or domains:

I. We have renounced consideration of the formation of bound nuclear clusters with $A > 2$, which may occur at very low density and/or at high (but not excessively high) temperature. The homogeneous unpaired (UP) phase is then always the ground state at high temperatures $T > T_{c0}$ in the restricted domain under study, where T_{c0} is the critical temperature of the pairing transition at $\alpha = 0$ (conventional).

II. The homogeneous isospin-asymmetric BCS phase is the ground state (denoted BCS) for all densities at intermediate temperatures (conventional).

III. The LOFF phase is the ground state in a narrow $T - \rho$ strip at low temperatures and high densities (unconventional).

IV. The domain of phase separation (PS), in either PS-BCS or PS-BEC realizations, appears at low temperatures (unconventional). In the joint PS-BCS phase, one of the components is the isospin-symmetric BCS phase, while the other is the normal isospin-asymmetric phase.

V. With decreasing density and intermediate or low temperatures, the phase diagram shows two types of BCS to BEC crossovers from the asymmetrical BCS phase to the BEC phase of deuterons and an embedded normal gas of excess neutrons: (i) A transition between the homogeneous BCS/BEC phases at moderate temperatures (nominally conventional, but with a gas of leftover neutrons in the low-density limit) and (ii) a transition between the heterogeneous PS-BCS/PS-BEC phases at low temperatures (unconventional). By convention, boundaries between BCS (or PS-BCS) and BEC (or PS-BEC) phases are decided by the change of sign of the mean chemical potential $\bar{\mu}$. In the phase diagram these appear as nearly vertical lines, insensitive to α, seen in the low-ρ-low-T corner of the phase diagram.

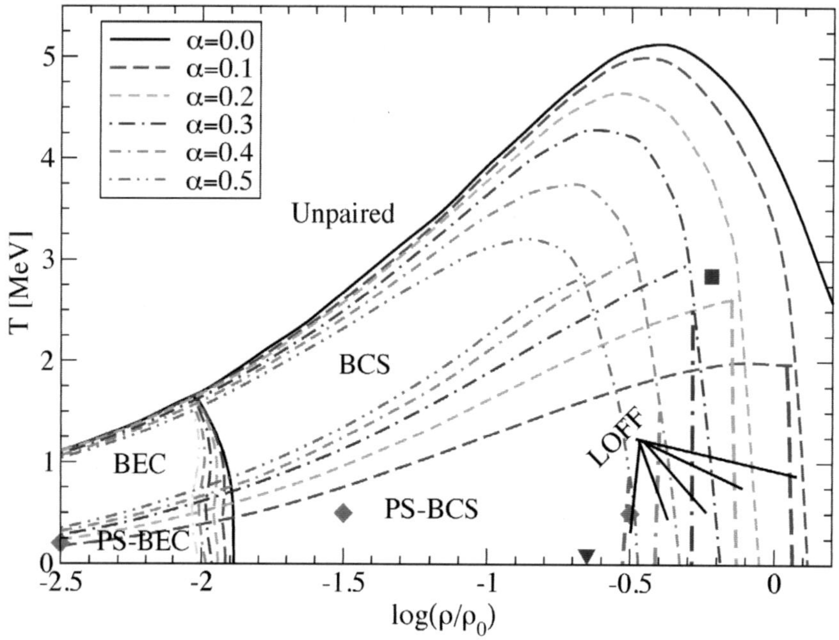

Figure 3. Phase diagram of dilute nuclear matter in the temperature-density plane at several isospin asymmetries α as indicated for the plotted lines. Four phases are represented: the normal, unpaired phase, the BCS (BEC) phase, the LOFF phase, and a PS-BCS (PS-BEC) phase. At each asymmetry there are two tri-critical points, one always being Lifshitz point. For special values of asymmetry these two tricritical points degenerate into single tetra-critical point, exemplified for $\alpha_4 = 0.255$ (marked by blue square). The LOFF phase disappears at the point $\log(\rho/\rho_0) = 0.65$ and $T = 0$ for $\alpha = 0.62$ (marked by blue triangle). Red triangles: representative points of strong, moderate, and weak coupling, left to right.

9. Special Features of the Phase Diagram

In terms of symmetries, four phases are identified: unpaired (UP), BCS/BEC, LOFF, and PS-BCS (PS-BEC). The results of the calculations performed are consistent with the transitions between BCS to BEC or between PS-BCS and PS-BEC being smooth crossovers without change of symmetry. The superfluid/unpaired phase transition and the transitions between superfluid phases are second order, except for those between PS-BCS and LOFF phases, which are of first order. At each nonzero isospin asymmetry $\alpha < 0.62$, the phase diagram exhibits two tricritical points where a simpler pairwise coexistence terminates and three different phases coexist, e.g., BCS, PS-BCS, and LOFF. One of these is always a Lifshitz point. For special values of asymmetry, the two tricritical points degenerate into a single tetracritical point; an example is shown as the blue square in Fig. 3, occurring at $\log(\rho/\rho_0) = -0.22$ and $T = 2.85$ for $\alpha = 0.255$.

10. Beyond the Phase Diagram

It is useful to distinguish three dynamical regimes:

- The weak-coupling regime (WCR) corresponds to the high-density limit where well-defined Cooper pairs are present.
- The strong-coupling regime (SCR) corresponds to the low-density limit where well-defined

deuterons are formed.

- In between: intermediate coupling regime (ICR).

The following properties of pairing and condensates in low-density nuclear matter have been examined in detail within the same calculational framework:

- Temperature and asymmetry dependence of the pairing gap, with comparison of BCS and LOFF phases.
- The kernel of the gap equation as the momentum space-wave function of the Cooper pair, with comparison of BCS and LOFF phases.
- Evolution of the Cooper-pair wave function from the WCR through the ICR to the SCR, i.e., evolution from BCS pairing to the BEC condensate of deuterons.
- Occupation numbers of neutrons and protons; their behaviors from the BCS phase in WCR, through ICR, and on to BEC in SCR; LOFF in WCR.
- Quasiparticle excitations: dispersion relations for quasiparticle spectra in the $^3S_1$-$^3D_1$ BCS condensate and in the LOFF phase; evolution of spectral branches from WCR through ICR to SCR.

We next turn to selected samples of results from these informative studies. In figures providing results in the WC, IC, and SC regimes, typical values have been selected at (T, ρ) pairs specified as follows. WCR: $T = 0.5$ MeV, $\log(\rho/\rho_0) = -0.5$. ICR: $T = 0.5$ MeV, $\log(\rho/\rho_0) = -1.5$. SCR: $T = 0.2$ MeV, $\log(\rho/\rho_0) = -2.5$.

10.1. Behavior of the Pairing Gap

Results have been obtained for the pairing gap at density $\rho = 0.1$ fm^{-1} (a) as a function of temperature for different asymmetry values, and (b) as a function of asymmetry for different temperatures (see Figs. 4 and 5). When the possibility of a LOFF phase is taken into account, these results, when plotted for each value of α, reveal different regimes at relatively low and relatively high temperature. The high-temperature segment corresponds to the BCS state, with standard temperature dependence of the gap. By contrast, in the low-temperature region below the branch point, there are two competing phases: BCS and LOFF, with very different temperature dependences of the gap function.

The quenching of the BCS gap upon decrease of T is caused by the loss of coherence among the quasiparticles as the thermal smearing of the Fermi surfaces is terminated, with the (unorthodox) consequence that for large enough asymmetries there exists a lower critical temperature $T_{c\downarrow}$.

In the plots of Fig. 5 showing Δ versus α for several chosen temperatures and fixed density, there are two curves for each T value: one in the low-α regime where only the BCS phase exists, and the other in the large-α regime where both BCS and LOFF are possible. The LOFF solution wins the competition in the latter region, since it provides larger gap values.

10.2. Kernel of the Gap Equation and Cooper-Pair Wave Function

The kernel of the gap equation,

$$K(k) = \frac{1}{2}\sum_{a,r} \frac{1}{2\sqrt{E_S^2(k) + \Delta^2(k,Q)}}[1 - 2f(E_a^r)], \tag{9}$$

is the product of the imaginary part of the retarded anomalous propagator and the Pauli operator $P_r^a = 1 - 2f(E_a^r)$. Physically, $K(k)$ can be interpreted as the momentum-space wave function of the Cooper pairs, since it obeys a Schrödinger-type eigenvalue equation in the limit of extremely strong coupling. The Pauli operator is a smooth function of momentum with a maximum at the

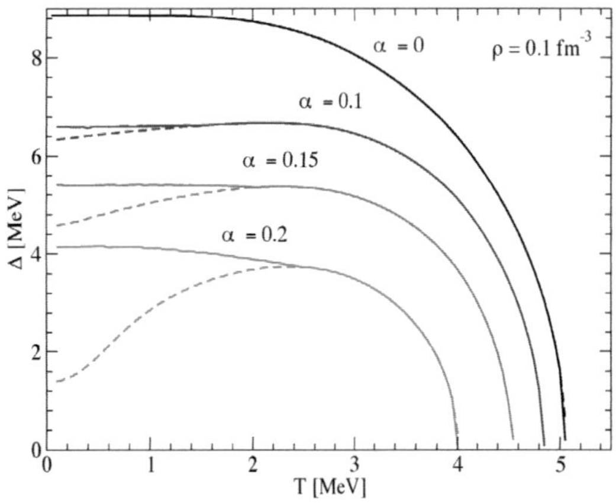

Figure 4. Gap as a function of temperature at constant density for asymmetry values $\alpha = 0.0$ (black), $\alpha = 0.1$ (blue), $\alpha = 0.15$ (red), $\alpha = 0.2$ (magenta). Results allowing for the LOFF phase are traced by solid lines, those restricted to the BCS phase, by dashed lines.

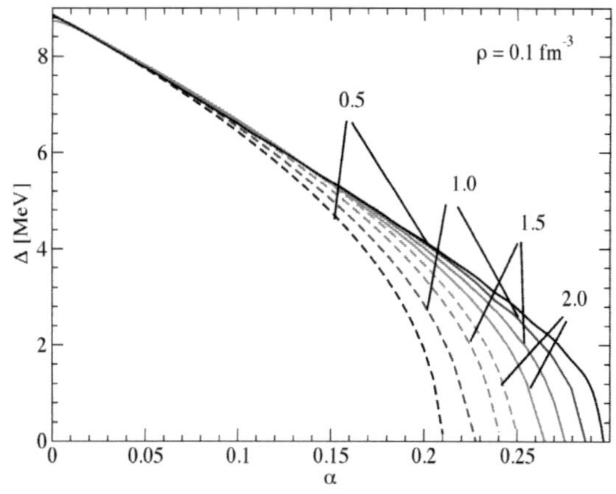

Figure 5. Gap as a function of asymmetry at constant density for temperature values $T = 0.5$ (black), $T = 1.0$ MeV (blue), $T = 1.5$ MeV (red), $T = 2.0$ MeV (magenta). Results allowing for the LOFF phase are traced by solid lines, those restricted to the BCS phase, by dashed lines.

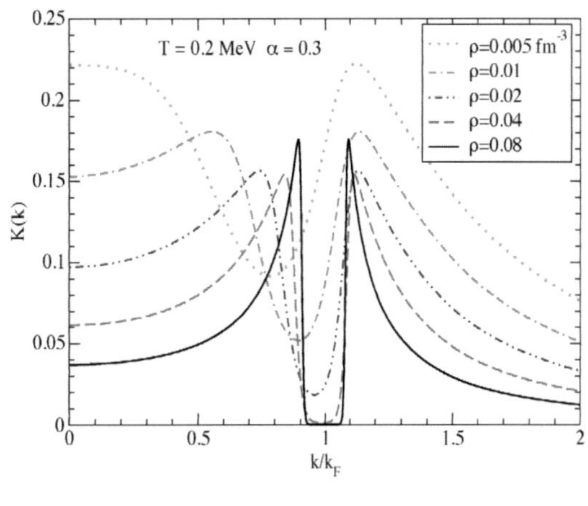

Figure 6. Dependence of the kernel $K(k)$ on momentum in units of Fermi momentum for fixed $T = 0.2$ MeV, $\alpha = 0.3$, and various densities indicated in the plot.

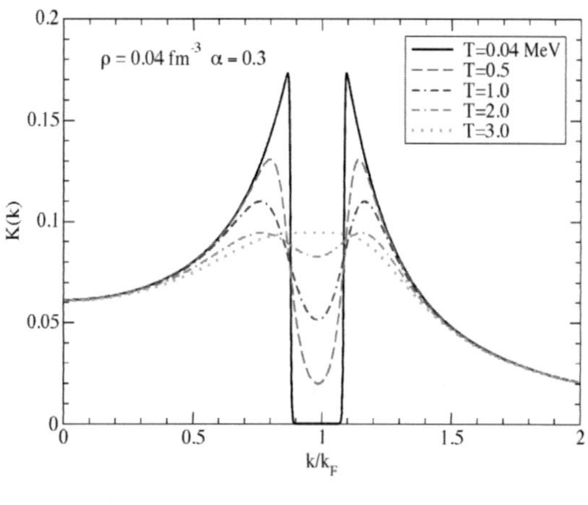

Figure 7. Dependence of the kernel $K(k)$ on momentum in units of Fermi momentum for fixed $\rho = 0.04$ fm^{-3}, $\alpha = 0.3$, and various temperatures indicated in the plot.

Fermi surface, where E_S vanishes. In practice, contributions from the two relevant excitation branches with $r \neq a$ are kept; those from the cases $r = a$ are negligible. Plots showing the momentum dependence of the kernel for asymmetry $\alpha = 0.3$ and relevant values of density and temperature are provided in Figs. 6 and 7.

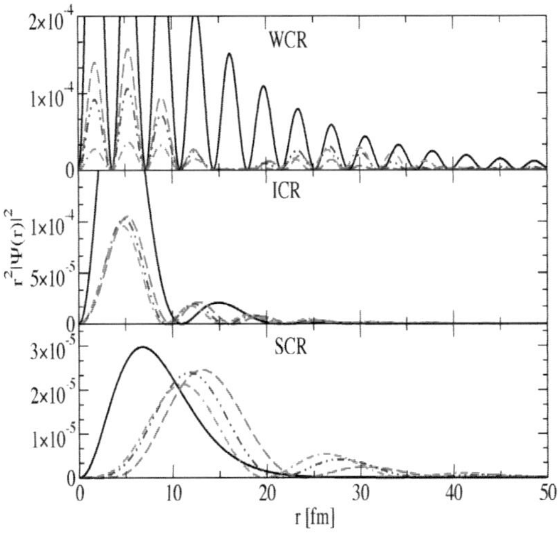

Figure 8. Dependence of $r^2|\Psi(r)|^2$ on r for the three coupling regimes. Color-coded α values: 0.0 (black), 0.1 (red), 0.2 (blue), 0.3 (magenta).

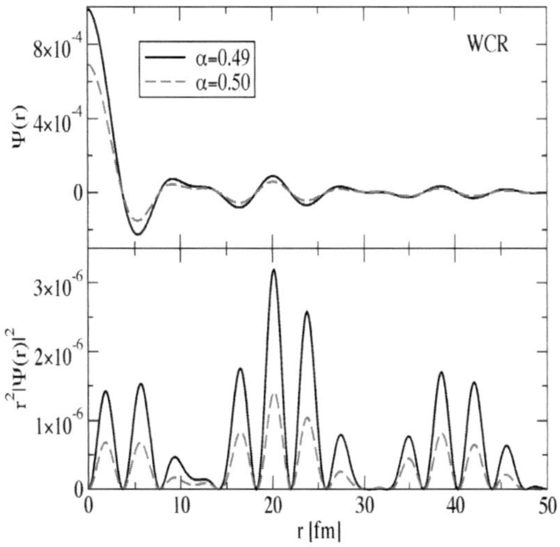

Figure 9. Dependence of $\Psi(r)$ and $r^2|\Psi(r)|^2$ on r in the WCR for two asymmetries at which the LOFF phase is the ground state.

10.3. Cooper-Pair Wave Function and Correlation Length

The superfluid coherence length ξ is directly related to the root-mean-square radius of the **r**-space Cooper-pair wave function, given by

$$\Psi(\mathbf{r}) = \mathcal{N} \int \frac{d^3p}{(2\pi)^3} [K(\mathbf{p}, \Delta) - K(\mathbf{p}, 0)] e^{i\mathbf{p}\cdot\mathbf{r}}, \tag{10}$$

where the factor $\mathcal{N}$ ensures unit norm for $\Psi(\mathbf{r})$. The mean-square radius of the Cooper pair is then defined as

$$\langle r^2 \rangle = \int d^3r r^2 |\Psi(\mathbf{r})|^2, \tag{11}$$

and its spatial extent, the coherence length, as $\xi_{\rm rms} = \sqrt{\langle r^2 \rangle}$. In traversing the BCS-BEC crossover, the change in the coherence length tracks the change of the condensate wave function. The regimes of weak and strong coupling can be identified by comparing the coherence length, given in the BCS case by the familiar result $\xi_a = \hbar^2 k_F / \pi m^* \Delta$, to the mean distance $d = (3/4\pi\rho)^{1/3}$. Detailed computation and analysis based on these relations firmly establishes that in the BCS limit (WCR) one is dealing with a coherent state whose wave function oscillates over many periods characterized by k_F^{-1}. In the opposite limit (SCR), the wave function is concentrated around the origin, indicating that one is dealing with a Bose condensate of strongly bound states: deuterons. Plots illustrating the behavior of $r^2|\Psi(\mathbf{r})|^2$ versus r in the different coupling regimes and for selected asymmetries α are provided in Figs. 8 and 9.

10.4. Occupancies, Excitations, and Pauli Blocking Effects

Analysis of the kernel $K(k)$, the p and n occupation probabilities, and quasiparticle dispersion relations reveal prominent effects of Pauli blocking (the "breach"). Summarizing the phenomena occurring in the different coupling regimes, we find

- WCR: At large asymmetries, the minority component is expelled from the blocking region ($n_p \approx 0$), while the majority component is maximally occupied ($n_n/2 \approx 1$). The "breach" is filled in with increasing T.

- WCR: The LOFF phase appearing in this regime largely mitigates the blocking mechanism by allowing for a non-zero CM momentum of the condensate. Accordingly, all intrinsic properties are much closer to those of the isospin-symmetric BCS phase.
- WCR: In the small-α limit, the occupation numbers are clearly fermionic, with some diffuseness due to the temperature.
- ICR: The fermionic nature of the occupation numbers is lost, a Fermi surface cannot be identified, and no "breach" appears.
- ICR: For large α values, the occupation numbers become non-monotonic; for the minority component this is a precursor of a change in the topology of the Fermi surface in the transition ICR $\to$ SCR.
- SCR: the occupation numbers and other properties are consistent with a BEC of strongly-coupled pairs (deuterons). At large asymmetries, the Fermi sphere of the minority component in the WCR has evolved into a shallow shell structure.
- SCR: long-range coherence of the condensate is lost.
- WCR $\to$ ICR $\to$ SCR: The quasiparticle dispersion relation changes in form from that corresponding to the existence of a Fermi surface to one that is minimal at $k = 0$, independent of isospin asymmetry.
- With increasing α, the proton component acquires points with zero excitation energy, as in gapless superconductivity. The occupation numbers reach a maximum at finite k and reflect a change of topology: from the filled Fermi sphere to one with an empty "core."

11. The Missing Ingredient: Formation of Nuclear Clusters

The matter in supernova envelopes is (i) at finite isospin asymmetry (though remaining quite small compared to its values in the crust of a neutron star), and (ii) at relatively high temperatures compared to those considered in the present study. This environment could support the existence of a substantial population of nuclear clusters besides deuterons, notably tritons, ^{3}He nuclei, and alpha particles. The α particles may form a BEC at sufficiently low temperatures. The extent to which the presence of such clusters will modify the structure of the phase diagram constructed so far remains to be determined at a comparable level of microscopic precision. Additionally, formation of neutron dimers in the nuclear medium remains a tantalizing possibility, which will be explored schematically in the second part of this review.

12. Framework for BCS Pairing Versus Hidden Dimer State: "Take One"

The NN interaction in the $^1S_0$ partial wave, having a scattering length $a = -18.95$ fm and a strong resonance at the tiny energy of 0.067 MeV, just barely fails to support a bound state. Although this hypothetical dineutron cannot exist in vacuum, there remains a tantalizing potential for it to be bound in a nuclear medium and even in pure neutron matter. For example, it would take an effective, in-medium mass M^* of about 5% over the bare mass M for a dineutron to bind in the pure neutron system. One could think of such a dineutron as a nuclear analog of the polaron of solid-state physics, which also has no existence in vacuum.

Recently, Eckhard Krotscheck's quantum many-body group at Buffalo [18] has developed and executed a powerful *ab initio* approach within correlated-basis theory that is capable of quantitative prediction of the ground-state properties of strongly interacting Fermi systems in the low-density regime, with errors in the ground-state energy below 2%. For two-body interactions qualitatively similar to the S-wave neutron-neutron potential, these authors have exploited rigorous variational aspects of this approach to demonstrate the existence of a robust singularity at low (but not asymptotically low) density that is plausibly associated dimerization of the system driven by phonon exchange. This behavior is analogous to what occurs at zero density when the in-vacuum scattering length diverges; accordingly, it is interpreted in terms of divergence of a well-defined in-medium scattering length at finite density. On a more phenomenological plane, it is important to highlight not only historical [23] early interest in possible existence of a dineutron, but more significantly an extensive body of recent theoretical and experimental work [24], respectively predicting and providing evidence for strong dineutron correlations in finite nuclei or nuclear matter, even suggesting a tendency toward Bose-Einstein condensation of long-lived dineutrons.

Our objective for this section of the review and the next is to furnish a schematic framework [5, 25] for the emergence of a hidden dimer state in homogeneous Fermi matter (potentially a dineutron in the case of realistic nuclear matter), which competes energetically with BCS pairing states, both conventional and unconventional. The present analysis follows the same paths as explored in [5], but uncovers additional subtleties beyond what is revealed in that work. In some important aspects, this treatment supplements the *ab initio* study carried out in [18] within the method of correlated basis functions. The analysis is primarily based on the Thouless criterion for determination of the critical temperature for the onset or termination of pairing correlations. The Thouless criterion [26, 16] centers on the behavior of the linear integral equation

$$\mathcal{T}(p,T) = -\int V(p,p_1)\frac{\tanh\left(\epsilon(p_1)/2T\right)}{2\epsilon(p_1)}\mathcal{T}(p_1,T)dv_1. \tag{12}$$

for the finite-temperature T-matrix (denoted here by $\mathcal{T}$), stating that if it does not have a pole at temperature T, this temperature lies above above the transition temperature. The single-particle spectrum appearing in this equation is given by $\epsilon(p) = p^2/2M - \mu$, with M the bare mass and μ the chemical potential.

Analysis of Eq. (12) is expedited by decomposing the momentum-space pairing interaction $V(p_1,p_2)$ identically into a separable component and a remainder $R(p_1,p_2)$ that vanishes when either momentum argument is on the Fermi surface:

$$V(p_1,p_2) = V_F\phi(p_1)\phi(p_2) + R(p_1,p_2). \tag{13}$$

The latter property is guaranteed by the choice $\phi(p) = V(p,p_F)/V_F$, with $V_F = V(p_F,p_F)$. Inserting this expression into Eq. (12), simple algebra leads to an equivalent set of two coupled equations. The first is a linear integral equation for the shape $\chi(p) = \mathcal{T}(p)/\mathcal{T}(p_F)$ of $\mathcal{T}$ in momentum space:

$$\chi(p) = \phi(p) - \int R(p,p_1)\frac{\tanh\left(\epsilon(p_1)/2T\right)}{2\epsilon(p_1)}\chi(p_1)dv_1, \tag{14}$$

while the second equation takes the form

$$-\frac{1}{V_F} = \int \phi(p)\frac{\tanh\left(\epsilon(p)/2T\right)}{2\epsilon(p)}\chi(p)dv. \tag{15}$$

Due to the imposed behavior of the remainder function $\mathcal{R}$, the description of singular behavior in the Cooper pairing channel is isolated in the second equation. The same tactic was invoked to develop an efficient and accurate procedure [27] for solving BCS-type gap equations in those cases where the pairing interaction contains a strong inner repulsion in coordinate space. Widely used versions of the NN interaction, such as the Reid soft-core and Argonne V18, are of this character.

Introducing the difference $\eta(p) = \chi(p) - \phi(p)$, the second equation may be rewritten as

$$-\frac{1}{V_F} = I_{11}(T) + \int \phi(p)\frac{\tanh\left(\epsilon(p)/2T\right)}{2\epsilon(p)}\eta(p)dv, \tag{16}$$

where

$$I_{11}(T) = \int \phi(p) \frac{\tanh(\epsilon(p)/2T)}{2\epsilon(p)} \phi(p) dv, \tag{17}$$

while $\eta(p)$ obeys an integral equation replacing (14), viz.

$$\eta(p,T) = -\int R(p,p_1) \frac{\tanh(\epsilon(p_1)/2T)}{2\epsilon(p_1)} \left(\phi(p_1) + \eta(p_1,T)\right) dv_1. \tag{18}$$

At the bifurcation point $T = 0$, the first term on the right side of Eq. (16), i.e., $I_{11}(T)$, diverges logarithmically, behaving essentially as $0.5N(0)\ln(\epsilon_c/T)$. Therefore a solution $T = 0$ exists only if the second term *also* diverges at this point.

To confirm that a compensating divergence occurs, we expand the function $\eta(p)$ in a basis formed by the eigenfunctions $\zeta_n(p)$ of the kernel $R(p,p';\rho)L(p,0)$ involving the propagator

$$L(p,\omega) = -(1 - 2n(p))/(\omega - 2\epsilon(p) - i\delta\,\mathrm{sgn}(p - p_F)). \tag{19}$$

The expansion is dominated by the contribution from the eigenfunction $\zeta_0(p)$ belonging to the lowest eigenvalue σ_0 of this kernel, which satisfies

$$\zeta_0(p) = -\sigma_0 \int R(p,p_1,\rho) \frac{1}{2|\epsilon(p_1)|} \zeta_0(p_1) dv_1. \tag{20}$$

Thus we write

$$\eta(p,T) = \eta_0(T)\zeta_0(p) + \vartheta(p), \tag{21}$$

where the remainder $\vartheta(p)$ vanishes on the Fermi surface. Inserting this expression into the integral equation (18) for $\eta(p,T)$ and collecting all terms having $\eta_0(p,T)$ as a factor on the left, we have

$$\eta_0(T)\left(\zeta_0(p) + \int R(p,p_1) \frac{\tan(\epsilon(p_1)/2T)}{2\epsilon(p_1)} \zeta_0(p_1)\, dv_1\right) = Z(p), \tag{22}$$

where

$$Z(p) = -\vartheta(p) - \int R(p,p_1) \frac{\tan(\epsilon(p_1)/2T)}{2\epsilon(p_1)} (\phi(p_1) + \vartheta(p_1))\, dv_1. \tag{23}$$

It is convenient to use the eigenvalue equation (20) for ζ_0 to rewrite the left side of Eq. (22) as

$$\eta_0(T)\left(\frac{\kappa}{\sigma_0}\zeta_0(p) + \int R(p,p_1)\mathcal{D}(p_1,T)\zeta_0(p_1) dv_1\right) = Z(p), \tag{24}$$

where

$$\mathcal{D}(p,T) = \frac{\tanh(\epsilon(p)/2T)}{2\epsilon(p)} - \frac{1}{2|\epsilon(p)|} \tag{25}$$

and $\kappa \equiv \sigma_0 - 1$. Next, both sides of this result are multiplied by the product $\zeta_0(p)/(2|\epsilon(p)|)$, and the momentum integration is performed. The operator R may be eliminated from (22) with the aid of (20) to obtain

$$\eta_0(T) = (\kappa + \gamma(T))^{-1} I_{10}/I_{00}, \tag{26}$$

where $I_{00} > 0$ is given by

$$I_{00} = \int \zeta_0(p) L(p,0) \zeta_0(p)\, dv, \tag{27}$$

while

$$\gamma(T) = I_{00}^{-1} \int \zeta_0(p) \left(\frac{\tanh(\epsilon(p)/2T)}{2\epsilon(p)} - \frac{1}{2|\epsilon(p)|}\right) \zeta_0(p)\, dv \tag{28}$$

and

$$I_{10} = \sigma_0 \int \zeta_0(p) \frac{1}{2|\epsilon(p)|} Z(p)\, dv. \tag{29}$$

Equation (26) is the main result of this analysis.

To demonstrate that the sign of $\gamma(T)$ is positive, as required, the integrand in Eq. (28) is rearranged to write

$$\gamma(T) \propto \int \zeta_0^2(p) \frac{\tanh(|\epsilon(p)|/2T) - 1}{2|\epsilon(p)|} dv. \tag{30}$$

Since $\zeta_0(p)$ vanishes at the Fermi surface like $\epsilon(p)$, we may conclude that $\gamma(T) = \gamma T^2$, where henceforth γ is a positive constant.

Considering next the integral I_{10}, the explicit form for $Z(p)$ is inserted into its integrand. Noting that that the terms involving the remainder ϑ essentially cancel one another, we may then take

$$I_{10} = \int \zeta_0(p) \frac{1}{2|\epsilon(p)|} \phi(p) dv. \tag{31}$$

Since $\gamma(T)$ vanishes as T^2 for $T \to 0$, we see from Eq. (26) that the coefficient $\eta_0(T)$ does in fact diverge together with the function $\eta(p)$ itself at the critical point where $\kappa = \sigma_0 - 1$ changes sign.

We are now ready to substitute the result (26) into the dispersion relation (16). Implementing an energetic cutoff $\epsilon_c \sim \epsilon_F^0$ in the result, we then arrive at

$$\frac{1}{2} \ln(\epsilon_c/T) = \frac{1}{\lambda} - \frac{\nu^2}{\kappa(\rho) + \gamma T^2}, \tag{32}$$

where it is to be understood that all minor corrections are included in the effective pairing constant denoted λ.

This analysis was performed with the case ($\lambda > 0$, $\kappa < 0$) in mind. Near the critical density ρ_t, the "stiffness" parameter κ, being small and negative, behaves as

$$\kappa \sim -|(\partial\kappa/\partial\rho)_t|(\rho_t - \rho). \tag{33}$$

For small $|\rho_t - \rho|$, the contribution of the $1/\lambda$ term on the right side of Eq. (32) can be neglected, and we find

$$T_1 \propto \exp[-(a\rho_t)/(\rho_t - \rho)], \tag{34}$$

where a is a numerical constant. Since the BCS constant λ does not enter this result, we infer that the solution so obtained does not belong to the BCS type.

However, there is in fact a second root of Eq. (32), lying at temperatures $T_2 > \sqrt{|\kappa|/\gamma} > T_1$. To assess its location, we set $\kappa = 0$ in Eq. (32) to write

$$\frac{\lambda\gamma}{2} T^2 \ln(\epsilon_c/T) = \gamma T^2 - \lambda\nu^2. \tag{35}$$

The solution of this equation has the form

$$T_2{}^2 = \lambda\nu^2/\gamma + O(\lambda^2 \ln\lambda). \tag{36}$$

Since the result (36) depends explicitly on the BCS constant λ, it is this solution that corresponds to Cooper pairing. We conclude that the dineutron state loses in its competition with the BCS-like state (36), as long as the magnitude of the critical parameter $|\kappa|$ remains small.

The situation at $T \to 0$ changes completely when $|\kappa|$ increases beyond $\lambda\nu^2$, for then the first term on the right side of Eq. (32) prevails over the second. As a result, the first root of

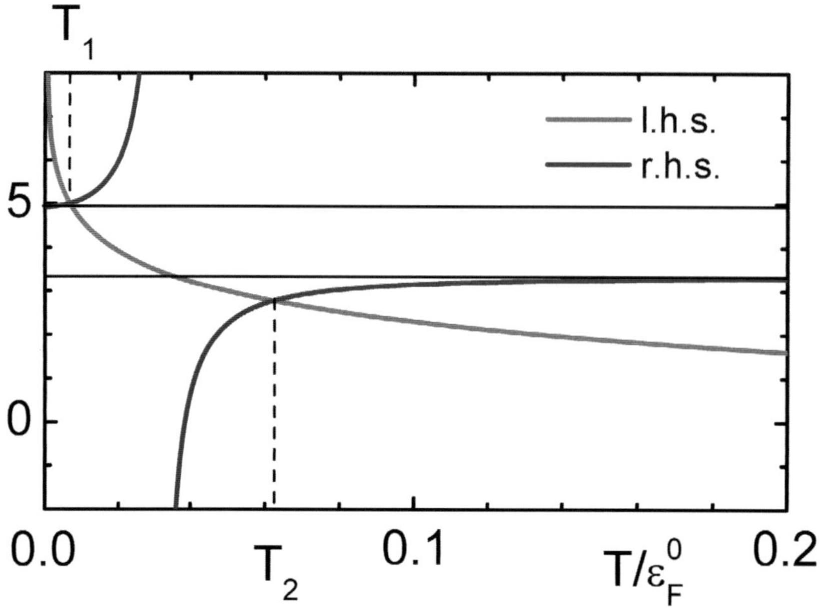

Figure 10. Graphical demonstration of two solutions of Eq. (32). Model parameters adopted: $\lambda = 0.3$, $\nu = 0.04$, $\kappa = -0.001$, $\gamma = (\epsilon_F^0)^{-1/2}$, $\epsilon_c = \epsilon_F^0$. The in-medium dineutron solution rather than the BCS solution ensures the maximum critical temperature for termination of pairing correlations provided $|\kappa| > \lambda\nu^2$; otherwise the BCS solution still holds.

this equation has the standard BCS form $T_1 = \epsilon_c \exp[-1/\lambda]$. However, as before there exists a different and larger root, determined by the relation

$$T_2 = T_{nn} = \sqrt{(|\kappa| + \lambda\nu^2)/\gamma} \simeq \sqrt{|\kappa|/\gamma}, \tag{37}$$

implying that the corresponding solution is of dineutron character. Thus, at sufficiently large $|\kappa|$ the dineutron state wins the contest for survival.

Fig. 10 shows a graphical solution of the key relation (32) derived above, the running variable T being measured in units of $\epsilon_F^0 = p_F^2/2M$. The left side of this equation, plotted versus T, is traced by the red line, while the right side is drawn as two disjoint blue lines. Model inputs are stated in the figure caption. The two crossing points determine two critical temperatures, T_1 and T_2, with $T_2 > T_1$.

13. Framework for BCS Pairing Versus Hidden Dimer State: "Take Two"

We now expand the analysis from the case ($\lambda > 0$, $\kappa < 0$) to the other quadrants of the $\lambda - \kappa$ parameter plane. The strategy now is to track the location of the Cooper singlet-channel pole of the zero-temperature scattering amplitude in the normal state of the homogeneous many-fermion system. We pursue this task based on the Bethe-Salpeter equation for the corresponding vertex part $\mathcal{K}_{\alpha\beta}(\mathbf{p}, \omega) = \mathcal{K}(\mathbf{p}, \omega)(\tau_2)_{\alpha\beta}$. Here α, β are spin indices, $\mathbf{p}$ is the momentum of the incoming particle (its target having momentum $-\mathbf{p}$), and ω is the total two-particle energy measured from 2μ. The equation ready for analysis now reads

$$\mathcal{K}(p, \omega) = -\int V(p, p_1) L(p_1, \omega) \mathcal{K}(p_1, \omega)\, dv_1, \tag{38}$$

where $V(p, p_1)$ represents the block of diagrams irreducible in the particle-particle channel (here just the pairing interaction) and $L(p, \omega)$, again given by Eq. (19), is the particle-particle propagator in the normal ground state. The analysis of the integral equation (38) now proceeds essentially in parallel with that carried out for Eq. (12). We bypass the lengthy details and move immediately to the final result for the basic dispersion equation, whose analytic continuation to the complex-ω plane yields

$$\frac{1}{2}\left(\ln\frac{\epsilon_c}{\omega} + i\frac{\pi}{2}\right) = \frac{1}{\lambda} - \frac{\nu^2}{\kappa + B(\omega)}, \tag{39}$$

where $\nu^2 = I_{10}^2/I_{00}N(0)$, with $N(0)$ denoting the normal density of states. The function $B(\omega)$ is given by

$$B(\omega) = -(I_{00})^{-1}\int \zeta_0(p)\delta L(p,\omega)\zeta_0(p)\, dv, \tag{40}$$

with $\delta L(p, \omega) = L(p, \omega) - L(p, 0)$. Setting $\Omega = i\omega$ in Eq. (39), we arrive at the general formula for determination of the pairing-gap proxy Ω,

$$\frac{1}{2}\ln\frac{\epsilon_c}{\Omega} = \frac{1}{\lambda} - \frac{\nu^2}{\kappa + B\Omega^2\ln(\epsilon_c/\Omega)} \tag{41}$$

with $B > 0$.

These results prepare us to uncover systematically the nature of pairing solutions in relevant domains of the $\lambda - \kappa$ plane, while acknowledging that in a *realistic* two-nucleon interaction these two parameters actually constrain one another. (We note that the quadrant ($\lambda < 0$, $\kappa > 0$) is empty of solutions since the right side of Eq. (41) is never positive.)

- *Quadrant* ($\lambda > 0$, $\kappa > 0$). The effect of the second term on the right side of Eq. (41), proportional to ν^2, amounts just to a renormalization of the coupling constant λ in the first term. Only the BCS solution survives in this case.
- *Quadrant* ($\lambda > 0$, $\kappa < 0$). This case has been dealt with earlier in the context of the Thouless criterion. The parameter κ goes negative at densities below some threshold density ρ_t, thereby triggering the onset of the dineutron state. As already established above, two different solutions exist in this quadrant, a BCS solution and another corresponding to a hidden dineutron state. It is informative to track the trajectories of both roots $\Omega_{1,2}$ as functions of λ. As λ goes to zero, the left root (nearest the origin) has the behavior $\Omega_1 \propto e^{-2/\lambda}$, with $\Omega_1(0) = 0$. This is obviously the ordinary BCS root. In the same limit, the other root, Ω_2, is situated close to $\sqrt{|\kappa|/B}$ and is clearly of non-BCS character. With increasing λ, both roots move away from the origin. It should be emphasized that in the analysis performed, both parameters λ and κ are assumed to remain small to assure the smallness of the roots, but with this proviso the analysis performed is self-consistent. As the roots evolve with increasing λ, the two solutions switch ascendancy in the region where $\lambda\nu^2 \simeq |\kappa|$. Thus we find that the dineutron state wins the competition with the BCS state only if the BCS coupling constant is rather small.
- *Quadrant* ($\lambda < 0$, $\kappa < 0$). Here only the root

$$T_{nn}^2 = (|\kappa| - |\lambda|\nu^2)/\gamma \tag{42}$$

survives, but it exists only in the limit of sufficiently large $|\kappa| > |\lambda|\nu^2$. Evidently, this solution is of dineutron character.

14. Conclusions and Outlook

In this mini-review we have discussed some recent (and not so recent) advances toward the goal of a comprehensive understanding of the phase diagram of nuclear matter in the low-density regime below about half the saturation density of heavy nuclei. In this density region one may safely regard neutrons and protons as the basic constituents, interacting through two-nucleon potentials constrained by nucleon-nucleon scattering data and binding energies of light nuclei (deuteron, triton, etc.). An *ab initio* many-body approach has been applied in a quantitative exploration of the temperature-density ($T - \rho$) phase diagram of this apparently well-defined quantum many-body system [4], focusing (a) on its dependence upon isospin asymmetry and (b) simultaneously on the roles played by pairing and the BCS-BEC crossover from Cooper pairs to bound bosonic dimers (deuterons). In addition to conventional BCS pairing, its competition with unconventional LOFF pairing (in which the Cooper pairs have finite total momentum) has been treated in some detail. Away from isospin symmetry, the heterogeneous phase which combines symmetrical superfluid and asymmetric normal matter may have lowest free energy. This possibility has been taken into consideration in the numerical calculations, with the finding that two-phase mixtures are favored over the homogeneous phases at the lowest temperatures and no-to-high density, where one of the components accommodates the excess neutrons.

The major portion of this review (Secs. 2-10) has dealt with the $T - \rho$ phase diagram of dilute nuclear matter in its dependence on isospin. In Secs. 12 and 13 the focus has been shifted to another intriguing aspect of the nuclear-matter problem. Restricting attention to pure neutron matter for simplicity, this is the possibility of unconventional pairing involving the emergence of neutron dimers, i.e., dineutrons, in some density range (obviously not including $\rho \to 0$). In turn, this eventuality raises the question of whether there is a corresponding BCS-BEC transition in pure neutron matter.

A very recent study carried out by some of us [19] has a bearing on these issues. Our initial study concentrated on the T-ρ phase diagram of neutron matter including only homogeneous and isotropic phases indexed by a spin-polarization parameter. The resulting diagram resembles that of asymmetrical nuclear matter because of the analogy between nuclear matter at non-zero isospin asymmetry and pure neutron matter at non-zero spin asymmetry. An essential distinction between the phase diagrams of these systems stems from the fact that two neutrons are not bound in vacuum, whereas a neutron and a proton bind into the deuteron in free space. Accordingly, pure neutron matter cannot, in the strict sense, undergo a BCS-BEC crossover from the ${}^1S_0$ BCS superfluid existing at higher densities, to a BEC condensate of neutron dimers at asymptotically low density. The end state must be a neutron gas. Nevertheless, the many-body calculations carried out recently by Stein et al. [19] give clear evidence of a BCS-BEC *precursor* and thus confirm earlier findings of various authors obtained at zero spin polarization (see [19] for references). The existence of a BCS-BEC precursor is revealed, for example, by the attainment of Cooper-pair correlation lengths comparable to the interparticle distance. This behavior must be attributed to the strongly resonant character of the nn interaction at low densities, where the ${}^1S_0$ component is dominant.

This discourse impels us to ponder the content of Secs. 12 and 13, which provide a schematic framework for understanding the emergence of unconventional types of pairing (notably, a "hidden dineutron state") along with the conventional BCS solution. A systematic classification of such solutions is achieved in terms of two parameters: a familiar coupling parameter λ and a so-called stiffness parameter κ. As introduced, these parameters would be responsible for subsuming the effects of such complications relative to basic, bare-interaction nuclear BCS theory as (i) medium-induced corrections to the in-vacuum pairing interaction (coming from exchange of density and spin-density fluctuations, for example) and (ii) non-trivial momentum dependence of the resulting dressed interaction. (Attention should also be given to self-energy corrections.) The first category of effects ("polarization corrections") is absent in the many-body treatment underlying the results presented in Secs. 2-10 and in [4, 19].

The recent, seminal work of Fan et al. [18] does consistently incorporate such effects in a quantitatively precise *ab initio* many-body approach developed within correlated basis theory. This approach has been exhaustively applied in a study correlations in the low-density, strongly interacting Fermi gas that addresses its Fermi-liquid state and BCS pairing, as well as dimerization induced by phonon exchange. Application of this method to neutron matter has the potential of resolving the fate of the dineutron, which may well depend (perhaps counterintuitively) on the hardness of the inner repulsion of the $^1S_0$ neutron-neutron interaction.

Acknowledgments
MS acknowledges support from the HGS-HIRe graduate program at Frankfurt University. AS is supported by the Deutsche Forschungsgemeinschaft (Grant No. SE 1836/3-1) and by the NewCompStar COST Action MP1304. XGH is supported by Fudan University Grant EZH1512519 and Shanghai Natural Science Foundation Grant No.14ZR1403000. Additionally, this research was partially supported by RFBR grants 13-02-00085, 14-02-00107, and 15-02-06261, and by grant NS-932.2014.2 from the Russian Ministry of Sciences. VAK thanks the McDonnell Center for the Space Sciences for timely support. JWC is indebted to the University of Madeira and its Centro de Ciências Matemáticas for gracious hospitality during his sabbatical residency.

References

[1] Sedrakian A and Clark JW 2006 *Phys. Rev.* C **73** 035803
[2] Stein M, Huang X-G, Sedrakian A, and Clark J W 2012 *Phys. Rev.* C **86** 062801(R)
[3] Stein M, Sedrakian A, Huang X-G, Clark J W, and Röpke G 2014 *J. Phys. Conf. Ser.* **496** 012008
[4] Stein M, Sedrakian A, Huang X-G, and Clark J W 2014 *Phys. Rev.* C **90** 065804
[5] Khodel V A, Clark J W, Shaginyan V R, and Zverev M V 2014 *Physics of Atomic Nuclei* **77** 1145
[6] Eagles D M 1969 *Phys. Rev.* **186** 456
[7] Nozières P and Schmitt-Rink S 1985 *J. Low Temp. Physics* **59** 195 1409.5649
[8] Alm T, Friman B L, Röpke G, and Schulz H 1993 *Nucl. Phys.* A **551** 45
[9] Baldo M, Lombardo U, and Schuck P 1995 *Phys. Rev.* C **52** 975
[10] Stein H, Schnell A, Alm T and Röpke G 1995 Zeits. f. Physik A **351** 295
[11] Lombardo U and Schuck P 2001 *Phys. Rev.* C **63**, 038201
[12] Mao S, Huang X, and Zhuang P 2009 *Phys. Rev.* C **79** 034304
[13] Heckel S, Schneider P P, and Sedrakian A 2009 *Phys. Rev.* C **80** 015805
[14] Huang X-G 2010 *Phys. Rev.* C **81** 034007
[15] Jin M, Urban M, and Schuck P 2010 *Phys. Rev.* C **82** 024911
[16] Ramanan S and Urban M 2013 *Phys. Rev.* C **88** 054315
[17] Sedrakian A 2013 *J. Phys. Conf. Ser.* **413** 012024
[18] Fan H H, Krotscheck E, Lichtenegger T, Mateo D, and Zillich R E 2015 *Phys. Rev.* A **92** 023640
[19] Stein M, Huang X-G, Sedrakian S, and Clark J W 2015 Spin-polarized neutron matter, critical pairing and BCS-BEC precursor *Preprint* nucl-th/1510.06000
[20] Fulde Peter and Ferrell R A 1964 *Phys. Rev.* **135** A 55; Larkin A I and Ovchinnikov Yu N 1964 Zh. Eksp. Teor. Fiz. 47: 1136
[21] Sedrakian A and Clark J W, in *Pairing in Fermionic Systems: Basic Concepts and Modern Applications* Eds Sedrakian A, Clark J W and Alford M (World Scientific, Singapore, 2006) 135
[22] Malfliet R A and Tjon J A 1969 *Nucl. Phys.* A **127** 161
[23] Feather N 1948 *Nature* **162** 213
[24] Ieki K et al 1996 *Phys. Rev.* C **54**, 1589; Matsuo M, Mizuyama K and Serizawa Y 2005 *Phys. Rev.* **71** 064326; Hagino K and Sagawa H 2005 *Phys. Rev.* C **72** 044321; Hagino K, Sagawa H, Carbonell J and Schuck P 2007 *Phys. Rev. Lett.* **99** 022506; Kanada-En'yo Y 2007 *Phys. Rev.* C **76** 044323; Isayev A A 2008 *Phys. Rev.* C **78** 014306; Papadimitriou G et al 2011 *Phys. Rev.* C **84** 051304(R); Sun T T, Sun B Y and Meng Ji 2012 *Phys. Rev.* C **86** 014305
[25] Khodel V A 1997 *Yad. Fiz.* **60** 1137
[26] Thouless D J 1960 *Ann. Phys.* (N.Y.) 553
[27] Khodel V A, Khodel V V, and Clark, J W 1996 **Nucl. Phys.** A **598** 390

Journal of Physics: Conference Series **702** (2016) 012013 doi:10.1088/1742-6596/702/1/012013

A further update on possible crises in nuclear-matter theory

W. H. Dickhoff

Department of Physics, Washington University, St. Louis, Missouri 63130, USA

E-mail: wimd@wuphys.wustl.edu

Abstract. The ancient problem of the saturation of symmetric nuclear matter is reviewed with an update on the status of the crises that were identified at an early stage by John Clark. We discuss how the initial problem with variational calculations providing more binding than the two hole-line contribution for the same interaction was overcome by calculations including three hole-line contributions without however reproducing the empirical nuclear saturation properties. It is argued that this remaining problem is still open because many solutions have been proposed or ad hoc adjustments implemented without generating universal agreement on the proper interpretation of the physics. The problem of nuclear saturation therefore persists leading to the necessity of an analysis of the way the nuclear saturation properties are obtained from experimental data. We clarify the role of short-range correlations and review results for nuclear saturation when such ingredients are completely taken into account using the Green's function method. The role of long-range correlations is then analyzed with special emphasis on the importance of attractive pion-dominated excitation modes which inevitably lead to higher saturation densities than observed. Because such modes have no counterpart in finite nuclear systems, it is therefore argued that they should not be considered when calculating nuclear matter properties. The remaining open question is then whether long-range correlations in finite nuclei which in turn have no counterpart in infinite matter, represent the remaining missing ingredient in this analysis. We also briefly comment on the role of three-body interactions in the context of the dispersive optical model description of experimental data. It is further noted that interactions based on chiral perturbation theory at present do not generate a sufficient number of high-momentum nucleons leading to radii that are too small and substantial overbinding in finite nuclei.

1. Introduction

On the occasion of the 80th birthday celebration of John Clark it is pertinent to recall his important contributions to the problem of the saturation properties of nuclear matter. For a more detailed summary of John Clark's contributions to many-body theory the laudatio for the award of the second Feenberg Medal is recommended reading [1]. Particularly relevant for the present discussion is his summary of the first many-body conference [2]. This meeting was dominated by presentations addressing aspects of the nuclear saturation issue. Variational methods had been developed [3] that demonstrated a substantial discrepancy with lowest-order Brueckner theory [4] employing the same interaction [5, 6]. The disagreement between different many-body techniques represented the first crisis identified by John Clark in Ref. [2]. This crisis made it clear that the next term in the so-called hole-line expansion [4, 7, 8] involving three independent hole lines could clarify and resolve this situation. The first implementation of this difficult calculation was accomplished by Day in Refs. [9, 10, 11]. Confirmation of the compatibility of different methods in establishing the equation of state was reported in Ref. [12]. Further clarification of the convergence of the hole-line expansion was developed by the Catania group [13] as will be discussed in more detail in Sec. 2.

The resolution of this first crisis implied that for nucleon-nucleon (NN) interactions with a substantial repulsive core it is possible to determine the equation of state under the assumption that nucleons are treated nonrelativistically while interacting only by means of two-body interactions. Added to these assumptions should be the notion that there is a limitation on the density range where these statements are valid. This limitation can be linked to the convergence criterion applied to the hole-line expansion and essentially translates into the notion that nuclear matter should still be a relatively low-density system [4]. Unfortunately, the resolution of the first crisis was followed by a second crisis as identified by John Clark [2] namely that nuclear-matter saturation properties were not explained in detail by the assumptions of nonrelativistic nucleons interacting by means of two-body interactions only. All results remained substantially removed from the empirical density of 0.16 nucleons per fm^3 and binding energy of 16 MeV per particle mostly in such a way that the binding energy is described correctly but at too high a saturation density. The latter systematics represent an improvement compared to the much more distinct Coester band obtained in lowest-order Brueckner theory for different interactions, where the location of the saturation point is related to the amount of D-state probability the interaction generates, large values leading to lower saturation densities and *vice versa* [14].

The resulting second crisis in nuclear-matter theory has been studied by many people and a multitude of remedies have been proposed. Since no uniform agreement on its resolution has emerged, this crisis endures to this day and several of its aspects will be discussed in the following. Section 2 is devoted to a discussion of some salient features of the three hole-line results of Ref. [13] together with some standard remedies for solving the discrepancy with the empirical saturation properties. Section 3 is devoted to a brief summary of what has been learned from Green's function Monte Carlo calculations in light nuclei [15] and recent developments that constrain the nucleon self-energy in ^{40}Ca [16, 17] including its high-momentum content. Insights from experiments that determine quantities relevant for saturation are discussed on the basis of the lessons provided by these considerations of finite nuclei. In particular, the value of the interior density appear to be dominated by the physics induced by short-range correlations (SRC). Other ingredients like the deviations of the mean field found in $(e, e'p)$ reactions are put together to provide an alternative perspective of nuclear saturation in Sec. 4. The impact of recent applications of NN interactions based on chiral perturbation theory is also briefly discussed. Some final words regarding John Clark's impact are presented in Sec. 5.

2. Ingredients of the hole-line expansion

The three hole-line calculation reported in Ref. [13] allowed for the first time an assessment of the choice of the auxiliary potential that can be introduced in the perturbation expansion of the ground-state energy. Many applications of lowest-order Brueckner theory had adopted the so-called standard choice in which the single-particle spectrum below the Fermi energy is calculated self-consistently from the Brueckner-Hartree-Fock (BHF) self-energy. The BHF self-energy can be calculated in analogy to the Hartree-Fock contribution but the bare interaction is replaced by Brueckner's G-matrix which sums all ladder diagrams while allowing only intermediate particle states corresponding to momenta above the Fermi momentum. In the standard choice the intermediate particle states in the G-matrix propagate with energies that correspond to kinetic ones reducing the computational effort. In the so-called continuous choice [18], the BHF prescription for the single-particle spectrum is extended to include momenta above the Fermi momentum as well and requires the recalculation of the G-matrix in each iteration step in which the single-particle spectrum is updated. In addition, only the real part of the BHF self-energy is

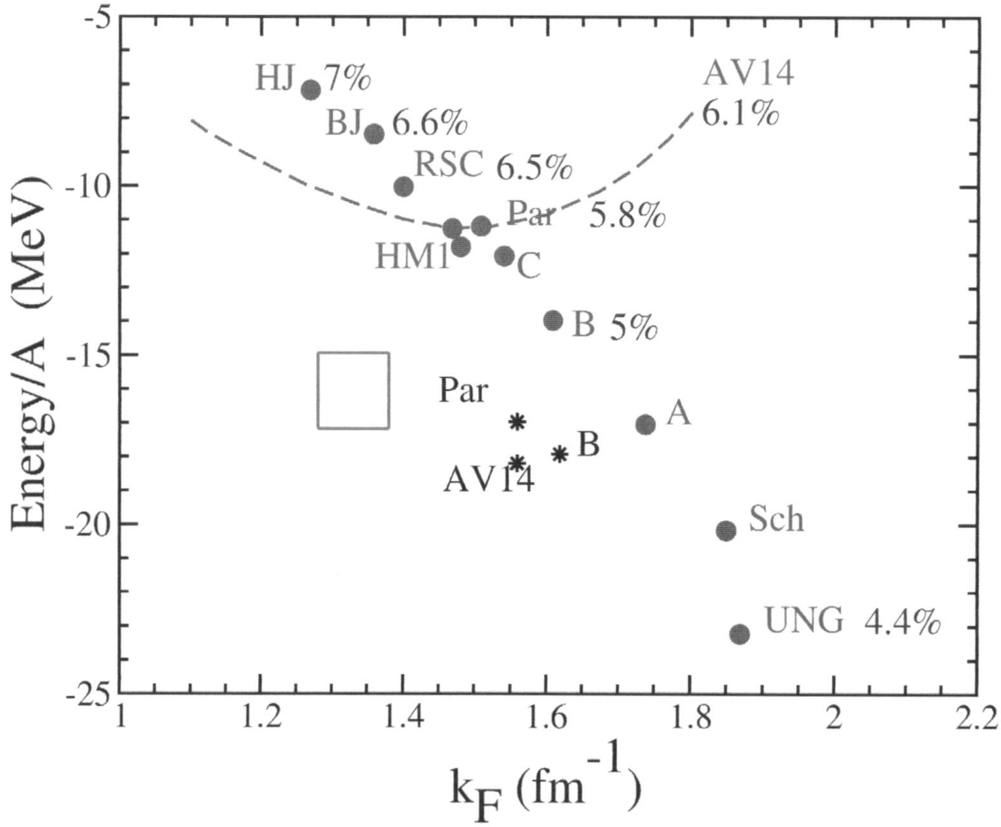

Figure 1. Energy per particle as a function of the Fermi wave vector for different realistic NN interactions. The filled circles indicate the minima of the saturation curve for the BHF approximation when the standard choice of U is made. The abbreviations identify the various interactions and in some cases the D-state admixture in the deuteron wave function is also given. The star symbols are associated with the minima of the saturation curve when three hole-line contributions are included. The box identifies the empirical region suggested by experimental data. For the Argonne $v14$ (AV14) interaction the saturation curve in the BHF approximation is represented by the dashed line. The figure has been adapted from Ref. [19].

included in this procedure (for momenta below the Fermi momentum there is no imaginary part). The three hole-line calculations of Ref. [13] included both choices for the auxiliary potential. As was well known at the time, the continuous choice provides consistently more binding than the standard choice at the level of two-hole lines. When three-hole line terms are added in each case the total result was almost indistinguishable at least for densities sufficiently beyond the expected saturation point demonstrating independence of the choice of the auxiliary potential and excellent convergence properties from two to three hole lines as suggested by Bethe. Furthermore, the continuous choice already approximated this final result reasonably at the two hole-line level allowing for a straightforward calculation of the equation of state at reasonable computational expense. Results summarizing the situation are illustrated in Fig. 1 which is adapted from Ref. [19]. The filled circles in Fig. 1 indicate the minimum of the energy per particle for different realistic NN interactions identified by the appropriate abbreviations. For the AV14 interaction [20] the dashed curve gives the relevant part of the complete curve. All open

Journal of Physics: Conference Series **702** (2016) 012013 doi:10.1088/1742-6596/702/1/012013

circles correspond to BHF calculations of the energy per particle confirming the identification of the so-called Coester band [14] with the location of the minimum governed by the D-state probability of the deuteron that is generated by the interaction. Including all three hole-line terms, moves the saturation points away from the Coester band towards the empirical region, as illustrated in Fig. 1 by the stars for three different realistic interactions [19]. The remaining discrepancy with empirical data is still substantial, however, especially when the results are plotted with respect to the density.

So the nuclear saturation problem remains! Several remedies have been proposed over the years and we will now consider some of their features. The first is closely associated with the presence of excited states of the nucleon, in particular the Δ-isobar. Its importance in pion-nucleon scattering suggests that it may be necessary to include it on the same footing as the nucleon. The disadvantage of this strategy is that a lot of information is required about the interaction between nucleons and Δ-isobars for which few experimental constraints are available. An alternative strategy is to represent the influence of Δ-isobars and other nucleonic excitations by including three- and perhaps higher-body interactions. The occurrence of such interactions is inevitable if one restricts the quantum Hilbert space to nucleons [21]. More elaborate versions of this type of three-body force [22] yield attractive contributions to the energy per particle. Since light nuclei require additional binding beyond the contribution of two-body interactions, such three-body forces help in getting better agreement for light nuclei. In nuclear matter, however, the situation is more complicated since the Coester band properties suggest that a repulsive mechanism is needed to generate lower saturation densities that are in accord with the empirical results. For this reason an additional phenomenological repulsive three-body interaction was introduced by [23], which was then adjusted to force the correct saturation properties of nuclear matter, while also fitting the binding of light nuclei. The procedure yields an improved Hamiltonian for nuclei but gives up on a deeper insight into the saturation mechanism of nuclear matter, especially since the origin of the effective repulsion is somewhat unclear. While the interactions have become more refined, the situation remains as described when two-body NN interactions are employed with strong repulsive cores.

An alternative solution to the saturation problem has been proposed that includes aspects of the effects of relativity [24]. A detailed discussion is beyond the scope of the present talk and only a few comments will be given here for completeness. By employing a straightforward adaptation of the BHF approach, the so-called Dirac–BHF (DBHF) method gives reasonable saturation properties for nuclear matter [25, 26]. The main physical effect appears to be the change of the coupling of the so-called σ-meson to the nucleons in the medium. This scalar, isoscalar meson represents to a large extent the physical exchange between nucleons of two interacting pions, coupled to zero angular momentum and isospin. Since the actual form of the scalar coupling of the σ-meson to the nucleons is essential in obtaining the saturation mechanism, it is unclear to what extent it truly represents the two pion-exchange processes in the medium. An additional difficulty is the necessity to deal with the properties of antiparticles and the so-called Dirac sea. Further study of higher-order (three hole-line) contributions have so far not taken place to assess the convergence properties of the scheme.

3. Lessons from finite nuclei

Some properties of light nuclei can nowadays be calculated in an exact manner with different techniques starting from a realistic NN interaction. An example is the application of several methods to the calculation of the ground-state energy of ^{4}He, reported in [27]. The low-lying states of nuclei up to $A = 10$ [28] can be described with the Green's function Monte Carlo (GFMC) method [15]. This body of work is able to explain many aspects of the low-energy spectra of light nuclei, starting from a realistic interaction. Many details are further improved by including a three-body interaction between the nucleons. In all cases studied

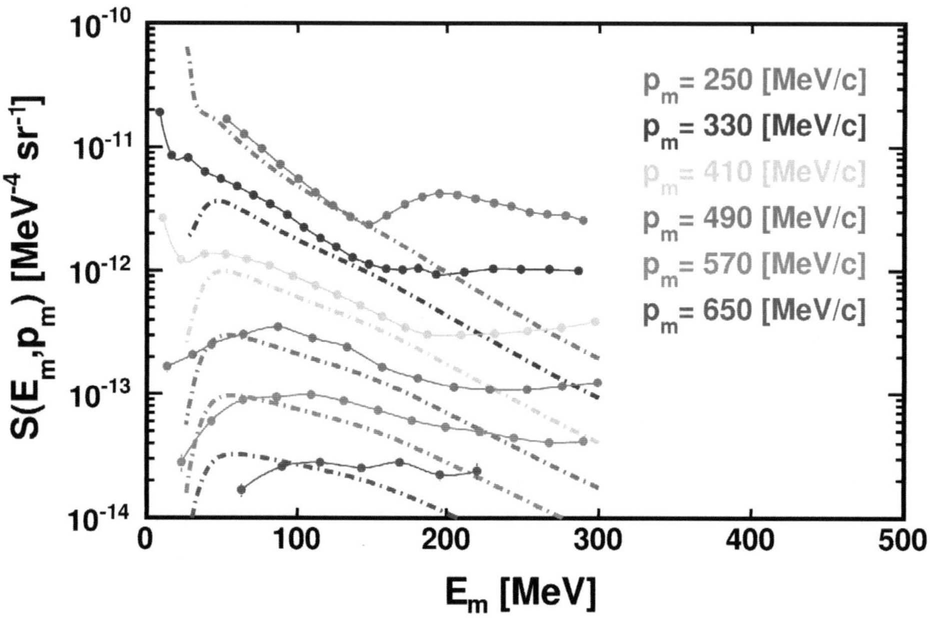

Figure 2. Spectral strength as a function missing energy for different missing momenta as indicated in the figure. The data are the average of the ^{27}Al and ^{56}Fe measurements from Ref. [34] and are represented by the filled circles.

so far, the calculated energy of the ground state is always above the experimental number when only two-body interactions are included: a clear indication for the need of an overall attractive three-body force corresponding to about 1.5 MeV per particle attraction for the Argonne calculations [15]. The two-body interaction used in the GFMC method is known as the AV18 [29]. It contains a substantial repulsive core capable of generating an appropriate amount of short-range correlations [30] demanded by exclusive electron scattering experiments [31]. With additional adjustments of the three-body interactions added to AV18 an excellent description of $p-$shell nuclei is obtained including the ground state of ^{12}C. We note that the AV18 is a local interaction which facilitates its implementation in the GFMC framework.

The issue of high-momentum components has recently been addressed in the context of developments of the dispersive optical model (DOM) originally developed by Mahaux and Sartor [32]. In Ref. [16] a nonlocal representation of the nucleon self-energy for ^{40}Ca was constrained by all available data related to the nucleon propagator including elastic scattering and structure information related to the ground state. The latter information includes removal of high-momentum nucleons obtained at Jefferson Lab [33, 34]. In Fig. 2 the results for the high-momentum removal spectral strength obtained in Ref. [16] are compared with the Jefferson Lab data [34]. The high-energy data correspond to intrinsic nucleon excitations and are not part of the DOM analysis. To further improve the description, one would have to introduce an energy dependence of the radial form factors for the potentials. Nevertheless an adequate description is generated which corresponds to 10.6% of the protons having momenta above 1.4 fm^{-1}.

The energy sum rule for the ground state [35] can be expressed as

$$E_0^N = \frac{1}{2\pi} \int_{-\infty}^{\varepsilon_F^-} dE \sum_{\alpha,\beta} \{\langle\alpha| T |\beta\rangle + E\, \delta_{\alpha,\beta}\} \text{ Im } G(\beta,\alpha;E), \tag{1}$$

where $G(\beta,\alpha;E)$ represents the single-particle propagator in a general basis. Equation (1) is valid when only two-body interactions are present in the Hamiltonian. In practice it is convenient to perform this calculation in momentum space employing the momentum distribution $n_{\ell j}(k)$ and $S_{\ell j}(k;E)$ the spectral function for a given ℓj combination. The results for ^{40}Ca quoted below are corrected for center-or-mass effects in the form given in Ref. [36]. A binding energy of 7.91 MeV/A is obtained much closer to the experimental 8.55 MeV/A than found in Ref. [37] for a local version of the DOM. The constrained presence of the high-momentum nucleons is responsible for this change [38]. The 7.91 MeV/A binding incorporates the contribution to the ground-state energy from two-body interactions including a kinetic energy of 22.64 MeV/A and was not part of the fit. This empirical approach therefore leaves about 0.64 MeV/A attraction for higher-body interactions but probably should be accompanied by an error of similar size at this stage. This statement requires consideration of the modification of the energy sum rule in the presence of *e.g.* a three-body interaction denoted by $\hat{W}$. The corresponding result can be expressed in the following form [39]

$$E_0^N = \frac{1}{2\pi} \int_{-\infty}^{\varepsilon_F^-} dE \sum_{\alpha,\beta} \{\langle\alpha| T |\beta\rangle + E\, \delta_{\alpha,\beta}\} \text{ Im } G(\beta,\alpha;E) - \frac{1}{2} \langle\Psi_0^N| \hat{W} |\Psi_0^N\rangle, \tag{2}$$

where the last term denotes the explicit contribution of the three-body term. The values discussed above have been obtained by employing Eq. (1) which in the presence of three-body interactions contains an implicit contribution from such terms. The form of Eq. (2) clarifies that in addition an explicit *repulsive* contribution of the three-body expectation value of 1.28 MeV/A is required on account of the extra minus sign for the $\hat{W}$ term in Eq. (2). This analysis can be compared to the approximately 1.5 MeV/A *attraction* needed for light nuclei in the GFMC calculations of Ref. [15] for light nuclei. The size of the three-body contribution whether attractive or repulsive of about 1.5 MeV/A suggests nevertheless that in comparison with about 30 MeV/A attraction from two-body terms there is a reasonable convergence with respect to the number of interacting bodies for the energy of the ground-state, assuming that the DOM spectral properties determining Eq. (1) are dominated by two-body contributions.

At this point it is helpful to recall how nuclear saturation properties are determined from experimental data. With this in mind we can consider whether and how these properties are determined by SRC and perhaps long-range correlations (LRC). For the latter consideration it is important to consider LRC in finite nuclei and infinite matter and determine if there is actually any relation between the two. Arguments are presented in the following that SRC represent the dominant factor in determining the empirical saturation density of nuclear matter. We recall that elastic electron scattering from ^{208}Pb [40] accurately determines the value of the central charge density. By multiplying this number by A/Z one obtains the relevant central density. For this nucleus one then finds 0.16 nucleons/fm^3 or $k_F = 1.33$ fm^{-1}. The presence of nucleons at the center of a heavy nucleus is confined to s-wave nucleons. Calculations of momentum distribution in nuclear matter suggest that the depletion of deeply bound states is dominated by SRC [41] generating a global depletion of the Fermi sea. The deviation from a constant depletion near the Fermi energy is due to the presence of LRC which only play a role in that energy domain.

Experimental confirmation of such features were provided by one of the last experiments at the NIKHEF facility employing the $(e,e'p)$ reaction on ^{208}Pb which covered the whole energy and momentum domain of the mean field for protons [42] (see also Ref. [35]). Experimental occupation numbers for mean-field proton orbits obtained in Ref. [42] demonstrate that also in the finite nucleus a global depletion of the Fermi sea occurs which exhibits only some orbital dependence near the Fermi energy. It is therefore particularly appropriate to conclude that the depletion of the deeply bound $0s\frac{1}{2}$ and $1s\frac{1}{2}$ protons is determined by SRC roughly corresponding to about 15% while it also pertains to a large extent for the $2s\frac{1}{2}$ orbit at the Fermi energy with only perhaps an additional 10% due to LRC [43]. These considerations demonstrate clearly that one may expect SRC to have a decisive influence on the actual value of the nuclear-matter saturation density extracted from finite nuclei. The above arguments can be applied to neutrons in ^{208}Pb and the A/Z multiplication may be more or less valid as there are four mostly occupied neutron s-orbits. Nevertheless an assumption about the nucleon asymmetry dependence of the saturation density will have to made as well. More unambiguous is the situation for ^{40}Ca where the interior charge density is expected to be accurately matched by the neutron one [16] on account of isospin symmetry. Accordingly, a value of about 0.16 nucleons per fm^3 does emerge in this nucleus.

The argument of the dominance of SRC was made for example in Ref. [44] and calculations to completely treat SRC in nuclear matter including full off-shell propagation in solving the ladder equation have been successfully implemented in particular at finite temperature [45, 46] using the self-consistent Green's function (SCGF) method [47]. As demonstrated in Ref. [44] for other realistic interactions, the latest nuclear-matter results for the AV18 interaction confirm that only treating SRC but fully self-consistently with results extrapolated to $T = 0$ generates saturation at the correct density [48]. The traditional amount of binding that is expected at saturation is obtained by extrapolating the empirical mass formula for $N = Z$ to infinite particle number while discarding the Coulomb contribution. The resulting value suggests a binding at saturation of about 16 MeV$/A$. The amount of binding at saturation for AV18 from the SCGF calculation reported in Ref. [48] is about 4 MeV short. This issue will be addressed in the next section.

4. Alternative perspective of nuclear saturation

It remains to be understood why apparently converged hole-line calculations [13] yield higher saturation densities. The three hole-line terms obtained in Ref. [13, 48] indicate reasonable convergence properties compared to the two hole-line contribution. One may therefore assume that these results provide an accurate representation of the energy per particle of nuclear matter as a function of density for the case of nonrelativistic nucleons and two-body forces. At this point it is useful to identify a hidden assumption when the nuclear matter problem is posed. It asserts that the influence of LRC in finite nuclei and nuclear matter are commensurate. It has been suggested in Ref. [44] that this underlying assumption is questionable. The argument is based on the special properties of the LRC associated with pion-exchange interactions. These attractive LRC contribute at transferred wave vectors of about 1 to 2 fm^{-1}. Ring diagram summations of such attractive interactions yield a coherent sum (all terms are attractive). For interactions different from the Coulomb one, the integral over this momentum variable $\boldsymbol{q}$ generates no contribution to the ground-state energy for small values on account of the $dq\ q^2$ term. The pion-exchange terms do not suffer this fate, since they occur at finite q, and are amplified due to momentum conservation in an infinite system. It is unclear whether these terms actually generate a similar physical consequence in finite nuclei, where momentum is not a good quantum number. The nuclear matter response with pion quantum numbers demonstrates that low-lying strength should accumulate. The lack of such collectivity with pionic quantum numbers in nuclei illustrates that the relation between nuclear matter and finite nuclei, at least for these degrees of freedom, is nonexistent. Indeed, experimental data exhibit no enhanced response of the pion channel over that of the rho [49], as demanded by nuclear matter calculations. Given this inconsistency, one can conclude that binding-energy contributions of long-range pion-exchange terms **do not** play the same role in finite nuclei. It is therefore necessary to excise them from nuclear-matter calculations to establish contact with finite nuclei. In other words, the original nuclear matter problem has been ill-posed and only the effects of SRC should be employed to connect the infinite with the finite system.

Three hole-line contributions do include such a third-order ring diagram characteristic of LRC. The corresponding effect of these LRC on nuclear saturation properties is sizable, as shown by the results for three- and four-body ring diagrams calculated in Ref. [50] (see also Refs. [10, 13]). The results of Ref. [50, 51] demonstrate that such ring-diagram terms are indeed dominated by attractive contributions involving pion quantum numbers propagating around the rings and increase substantially in importance with increasing density. As has been argued above, these contributions have no counterpart in finite nuclei as they are simply too small. Excising the three ring-diagram contribution from the three hole-line result [13] indeed confirms that a sensible saturation density is obtained.

It is also important to note that even with the right nuclear-matter saturation properties nothing is explained if corresponding calculations in finite nuclei generate a nuclear charge density in the interior that is too large. Furthermore, it is in turn quite clear that LRC which clearly exist in finite nuclei as surface vibrations and giant resonances [52] have no corresponding counterpart in infinite nuclear matter. Whether such phonon contributions to the binding energy of finite nuclei are completely negligible is also not so obvious but remains an open question. It is certainly possible that such terms could contribute a few MeV per particle necessary to bridge the missing binding that apparently is generated in nuclear matter when only SRC are (sophisticatedly) included in the SCGF results for AV18 reported in Ref. [48].

We conclude this section with some brief remarks about the use of NN interactions based on chiral perturbation theory. In practical developments these interactions are cut in momentum space corresponding to about 500 MeV. If such a cut-off is introduced from the beginning, it is clear that no sizable SRC will be introduced. Indeed the momentum properties of nucleons in the ground state of nuclei observed experimentally at Jefferson Lab suggest [33, 34, 31, 53] that an interaction like AV18 is a much more appropriate vehicle to describe these nucleons [30]. It is therefore not surprising that *ab initio* calculations employing soft chiral interactions generate heavier nuclei that have a too small radius and exhibit substantial overbinding (see *e.g.* Ref. [54]). The impossibility to simultaneously to describe nuclear saturation and the triton binding energy with standard versions of chiral interactions is also noteworthy in this context [55]. The need for quite large corrections from three-body forces also appears inconsistent with the GFMC results for light nuclei and the empirical DOM result for ^{40}Ca [16]. Finally, it should be observed that lattice calculations that aim to generate the NN interaction directly from QCD [56] generate a strongly repulsive core when the pion mass assumes realistic values, thereby providing further indications that future applications to nuclear saturation and finite nuclei should proceed on the basis of NN interactions with strongly repulsive cores.

5. Dedication

The final words in this contribution should emphasize the importance of the efforts of John Clark in understanding nuclear-matter saturation. They continue to inspire some people like the author to persist in thinking about solutions to this problem in as complete and physically sensible a way as possible. Some of these ideas have been put forward in the present paper which is dedicated to John on the occasion of his 80th birthday!

Acknowledgment

This work was supported by the U.S. National Science Foundation under grants PHY-1304242.

References

[1] Bishop R F 1988 *Recent Progress in Many-body Theories* vol 1 ed Kallio A J, Pajanne E and Bishop R F (Plenum Press, New York) p 385
[2] Clark J W 1979 *Nucl. Phys. A* **328** 587
[3] Bäckman S O, Chakkalakal D A and Clark J W 1969 *Nucl. Phys. A* **130** 635
[4] Day B D 1967 *Rev. Mod. Phys.* **39** 719
[5] Clark J W 1979 *Prog. Part. Nucl. Phys.* **2** 89
[6] Pandharipande V R and Wiringa R B 1979 *Rev. Mod. Phys.* **51** 821
[7] Rajaraman R and Bethe H A 1967 *Rev. Mod. Phys.* **39** 745
[8] Bethe H A 1971 *Ann. Rev. Nucl. Sci.* **21** 93
[9] Day B D 1978 *Rev. Mod. Phys.* **50** 495
[10] Day B D 1981 *Phys. Rev. C* **24** 1203
[11] Day B D 1981 *Phys. Rev. Lett.* **47** 226
[12] Day B D and Wiringa R B 1985 *Phys. Rev. C* **32** 1057
[13] Song H Q, Baldo M, Giansiracusa G and Lombardo U 1998 *Phys. Rev. Lett.* **81** 1584
[14] Coester F, Cohen S, Day B D and Vincent C M 1970 *Phys. Rev. C* **1** 769
[15] Pieper S C and Wiringa R B 2001 *Annu. Rev.Nucl. Part. Sci.* **51** 53
[16] Mahzoon M H, Charity R J, Dickhoff W H, Dussan H and Waldecker S J 2014 *Phys. Rev. Lett.* **112** 162503
[17] Dussan H, Mahzoon M H, Charity R J, Dickhoff W H and Polls A 2014 *Phys. Rev. C* **90** 061603
[18] Jeukenne J P, Lejeune A and Mahaux C 1976 *Phys. Rep.* **25** 83
[19] Baldo M 1999 *Int. Rev. Nucl. Phys.* **8** 1
[20] Wiringa R B, Smith R A and Ainsworth T L 1984 *Phys. Rev. C* **29** 1207
[21] Fujita J and Miyazawa H 1957 *Prog. Theor. Phys.* **17** 360
[22] Coon S A *et al.* 1979 *Nucl. Phys. A* **317** 242
[23] Carlson J, Pandharipande V R and Wiringa R B 1983 *Nucl. Phys. A* **401** 59
[24] Anastasio M R, Celenza L, Pong W and Shakin C M 1983 *Phys. Rep.* **100** 327
[25] ter Haar B and Malfliet R 1986 *Phys. Rev. Lett.* **56** 1237
[26] Brockmann R and Machleidt R 1990 *Phys. Rev. C* **42** 1965
[27] Kamada K *et al.* 2001 *Phys. Rev. C* **64** 044001
[28] Pieper S C, Varga K and Wiringa R B 2002 *Phys. Rev. C* **66** 044310
[29] Wiringa R B, Stoks V G J and Schiavilla R 1995 *Phys. Rev. C* **51** 38
[30] Wiringa R B, Schiavilla R, Pieper S C and Carlson J 2014 *Phys. Rev. C* **89** 024305
[31] Subedi R *et al.* 2008 *Science* **320** 1476
[32] Mahaux C and Sartor R 1991 *Adv. Nucl. Phys.* **20** 1
[33] Rohe D *et al.* 2004 *Phys. Rev. Lett.* **93** 182501
[34] Rohe D 2004 *Habilitationsschrift* (University of Basel)
[35] Dickhoff W H and Van Neck D 2008 *Many-Body Theory Exposed!, 2nd edition* (New Jersey: World Scientific)
[36] Dieperink A E L and de Forest Jr T 1974 *Phys. Rev. C* **10** 543
[37] Dickhoff W H, Van Neck D, Waldecker S J, Charity R J and Sobotka L G 2010 *Phys. Rev. C* **82** 054306
[38] Müther H, Polls A and Dickhoff W H 1995 *Phys. Rev. C* **51** 3040
[39] Carbone A, Cipollone A, Barbieri C, Rios A and Polls A 2013 *Phys. Rev. C* **88** 054326
[40] Frois B *et al.* 1977 *Phys. Rev. Lett.* **38** 152
[41] Vonderfecht B, Dickhoff W H, Polls A and Ramos A 1991 *Phys. Rev. C* **44** R1265
[42] van Batenburg M F 2001 *Ph.D. Thesis* (University of Utrecht)
[43] Dickhoff W H 2010 *J. Phys. G* **37** 064007
[44] Dewulf Y, Dickhoff W H, Van Neck D, Stoddard E R and Waroquier M 2003 *Phys. Rev. Lett.* **90** 152501
[45] Frick T, Müther H, Rios A, Polls A and Ramos A 2005 *Phys. Rev. C* **71** 014313
[46] Rios A, Polls A and Dickhoff W H 2009 *Phys. Rev. C* **79** 064308
[47] Dickhoff W H and Barbieri C 2004 *Prog. Part. Nucl. Phys.* **52** 377
[48] Baldo M, Polls A, Rios A, Schulze H and Vidaña I 2012 *Phys. Rev. C* **86** 064001
[49] Carey T A *et al.* 1984 *Phys. Rev. Lett.* **53** 144
[50] Dickhoff W H, Faessler A and Müther H 1982 *Nucl. Phys.* **A389** 492
[51] Dickhoff W H 1984 *Prog. Part. Nucl. Phys.* **11** 529
[52] Brand M G E, Allaart K and Dickhoff W H 1990 *Nucl. Phys.* **A509** 1
[53] Arrington J, Higinbotham D, Rosner G and Sargsian M 2012 *Prog. Part. Nucl. Phys.* **67** 898
[54] Somà V, Cipollone A, Barbieri C, Navrátil P and Duguet T 2014 *Phys. Rev. C* **89** 061301(R)
[55] Hagen G, Papenbrock T, Ekström A, Wendt K A, Baardsen G, Gandolfi S, Hjorth-Jensen M and Horowitz C J 2014 *Phys. Rev. C* **89** 014319
[56] Ishii N, Aoki S and Hatsuda T 2007 *Phys. Rev. Lett.* **99** 022001

Journal of Physics: Conference Series **702** (2016) 012014 doi:10.1088/1742-6596/702/1/012014

Neutron matter with Quantum Monte Carlo: chiral 3N forces and static response

M Buraczynski[1], **S Gandolfi**[2], **A Gezerlis**[1], **A Schwenk**[3,4] **and I Tews**[3,4]
[1]Department of Physics, University of Guelph, Guelph, ON N1G 2W1, Canada
[2]Theoretical Division, Los Alamos National Laboratory, Los Alamos, NM 87545, USA
[3]Institut für Kernphysik, Technische Universität Darmstadt, 64289 Darmstadt, Germany
[4]ExtreMe Matter Institute EMMI, GSI Helmholtzzentrum für Schwerionenforschung GmbH, 64291 Darmstadt, Germany

Abstract. Neutron matter is related to the physics of neutron stars and that of neutron-rich nuclei. Quantum Monte Carlo (QMC) methods offer a unique way of solving the many-body problem non-perturbatively, providing feedback on features of nuclear interactions and addressing scenarios that are inaccessible to other approaches. In this contribution we go over two recent accomplishments in the theory of neutron matter: a) the fusing of QMC with chiral effective field theory interactions, focusing on local chiral 3N forces, and b) the first attempt to find an *ab initio* solution to the problem of static response.

1. Introduction

Neutron matter (NM) is an extended system composed of strongly interacting particles [1, 2]. Not only is it not a toy problem, but it provides crucial information for nuclear astrophysics and nuclear structure. Specifically, the neutron-matter equation of state (EOS) is directly relevant to neutron-star physics: the mass-radius relationship follows from using the EOS and general relativity. Furthermore, neutron matter offers significant input to energy-density functionals (EDFs) tailored to describe nuclei: in addition to fitting properties of known nuclei, the use of "synthetic data" (dependable *ab initio* many-body results) like the EOS of neutron matter extends the physics content of EDFs to the neutron-rich side. Since neutron matter is not self-bound, it provides an intriguing combination of features: it is strongly interacting but also easier to address than isospin-asymmetric systems and symmetric nuclear matter. This has led to the use of NM to probe detailed features of nuclear forces, of both phenomenological [3, 4, 5, 6, 7] and chiral varieties [8, 9, 10, 11, 12, 13, 14]. (See Ref. [15] for a mixed-potential approach).

The state-of-the-art in nuclear physics has advanced to the stage where several challenging problems are attacked using *ab initio* methods, i.e. from first principles: solving the many-nucleon Schrödinger equation. A particularly successful way of doing so is via the use of the non-perturbative many-body methods known as Quantum Monte Carlo (QMC). The two most successful continuum QMC methods used to describe nuclear systems are Green's Function Monte Carlo (GFMC) [16, 17, 18] and Auxiliary-Field Diffusion Monte carlo (AFDMC) [19]. Both methods follow the general philosophy of Diffusion Monte Carlo, extracting the ground state by starting out with a trial/variational wave function $|\Psi_T\rangle$ and projecting out excited states in imaginary time τ: $|\Psi_0\rangle = \lim_{\tau\to\infty} \exp(-\hat{H}\tau)|\Psi_T\rangle$. What makes the nuclear many-body problem especially challenging is that the interactions include central, spin, tensor, spin-

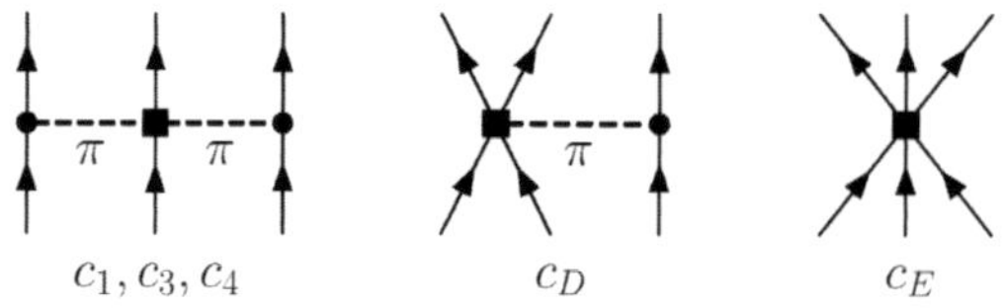

Figure 1. Three-nucleon forces at next-to-next-to-leading order. There are three distinct contributions: a) the two-pion-exchange (containing the couplings c_1, c_3, and c_4), b) the one-pion-exchange–contact interaction (containing c_D), and c) the contact interaction (containing c_E).

orbit, and more complicated terms. Both GFMC and AFDMC address the coordinate degrees of freedom stochastically. What distinguishes them is how they treat the spin-isospin degrees of freedom: in GFMC they are treated explicitly, while in AFDMC they are also handled stochastically, at the cost of introducing auxiliary fields. Since GFMC is limited to small sizes, AFDMC is more appropriate to describe extended systems like neutron matter, so in what follows we will focus on AFDMC results.

2. Chiral 3N forces

2.1. Formalism

Chiral Effective Field Theory (EFT) forces were developed in order to connect with the symmetries of Quantum Chromodynamics [20, 21, 22]. They contain pion exchanges and shorter-range contact terms. Specifically, they predict consistent three-nucleon (3N) forces, which first appear at next-to-next-to-leading order (N^2LO) [23, 24]. Chiral EFT interactions were adopted as input to many-nucleon calculations by the overwhelming majority of practitioners, with a notable exception: Quantum Monte Carlo methods typically require local potentials as input and were therefore unable to use chiral EFT forces. (For the sake of completeness, note that Monte Carlo methods have been used to study neutron matter based on lattice and other momentum-space techniques [25, 26, 27].) Chiral EFT interactions were usually non-local, due to having been constructed in momentum space (without regard to their locality or non-locality). Starting with Ref. [10], the non-locality of chiral EFT nucleon-nucleon interactions up to N^2LO was fully removed, arriving at local chiral forces that are fully equivalent to the older (momentum-space) potentials. These were then published and used in AFDMC for neutron matter [13], while also being implemented in GFMC for light nuclei [28].

The natural next step is to derive (and implement in QMC methods) the consistent three-nucleon interactions up to N^2LO. The relevant terms are shown in Fig. 1. Since QMC is naturally formulated in coordinate space, these contributions need to be Fourier transformed [29]. The 3N contact contribution V_E has the form:

$$V_E^{ijk} = \frac{c_E}{2f_\pi^4\Lambda_\chi}\sum_{\pi(ijk)} \boldsymbol{\tau}_i\cdot\boldsymbol{\tau}_k\,\delta(\mathbf{r}_{ij})\delta(\mathbf{r}_{kj})\,. \tag{1}$$

This contains a two-particle isospin term ($\boldsymbol{\tau}\cdot\boldsymbol{\tau}$) which is 1 in neutron matter; below we similarly find spin terms (e.g. $\boldsymbol{\sigma}\cdot\boldsymbol{\sigma}$). The one-pion-exchange–contact 3N interaction V_D has the following form:

$$V_D^{ijk} = \frac{c_D g_A}{24f_\pi^4\Lambda_\chi}\sum_{\pi(ijk)} \boldsymbol{\tau}_i\cdot\boldsymbol{\tau}_k\left[\frac{m_\pi^2}{4\pi}\,\delta(\mathbf{r}_{ij})X_{ik}(\mathbf{r}_{kj}) - \boldsymbol{\sigma}_i\cdot\boldsymbol{\sigma}_k\,\delta(\mathbf{r}_{ij})\delta(\mathbf{r}_{kj})\right] \tag{2}$$

which clearly contains not only a one-pion-exchange–contact part ($X\delta$), but also a contact–contact part ($\delta\delta$). Here and above, all the terms that are not delta functions should be taken to be long-range terms, with the details in Ref. [29]. The two-pion-exchange contribution V_C is most naturally given in three separate expressions:

$$V^{ijk}_{C,c_1} = \frac{c_1 m_\pi^4 g_A^2}{2f_\pi^4(4\pi)^2} \sum_{\pi(ijk)} \boldsymbol{\tau}_i \cdot \boldsymbol{\tau}_k \, \boldsymbol{\sigma}_i \cdot \hat{\mathbf{r}}_{ij} \, \boldsymbol{\sigma}_k \cdot \hat{\mathbf{r}}_{kj} \times U(r_{ij})Y(r_{ij})U(r_{kj})Y(r_{kj}) \tag{3}$$

and

$$V^{ijk}_{C,c_3} = \frac{c_3 g_A^2}{36 f_\pi^4} \sum_{\pi(ijk)} \boldsymbol{\tau}_i \cdot \boldsymbol{\tau}_k \times \left[\frac{m_\pi^4}{(4\pi)^2} X_{ij}(\mathbf{r}_{ij}) X_{kj}(\mathbf{r}_{kj}) \right.$$
$$\left. - \frac{m_\pi^2}{4\pi} X_{ik}(\mathbf{r}_{ij})\delta(\mathbf{r}_{kj}) - \frac{m_\pi^2}{4\pi} X_{ik}(\mathbf{r}_{kj})\delta(\mathbf{r}_{ij}) + \boldsymbol{\sigma}_i \cdot \boldsymbol{\sigma}_k \delta(\mathbf{r}_{ij})\, \delta(\mathbf{r}_{kj}) \right] \tag{4}$$

and

$$V^{ijk}_{C,c_4} = \frac{c_4 g_A^2}{72 f_\pi^4} \sum_{\pi(ijk)} \boldsymbol{\tau}_i \cdot (\boldsymbol{\tau}_k \times \boldsymbol{\tau}_j)$$
$$\times \left[\frac{m_\pi^4}{2i(4\pi)^2} [X_{ij}(\mathbf{r}_{ij}), X_{kj}(\mathbf{r}_{kj})] - \frac{m_\pi^2}{4\pi} \boldsymbol{\sigma}_i \cdot (\boldsymbol{\sigma}_k \times \boldsymbol{\sigma}_j)(1 - T(r_{ij}))Y(r_{ij})\delta(\mathbf{r}_{kj}) \right.$$
$$- \frac{m_\pi^2}{4\pi} \boldsymbol{\sigma}_i \cdot (\boldsymbol{\sigma}_k \times \boldsymbol{\sigma}_j)\,(1 - T(r_{kj}))\,Y(r_{kj})\delta(\mathbf{r}_{ij})$$
$$- \frac{3m_\pi^2}{4\pi} \boldsymbol{\sigma}_i \cdot \hat{\mathbf{r}}_{ij}\, \hat{\mathbf{r}}_{ij} \cdot (\boldsymbol{\sigma}_k \times \boldsymbol{\sigma}_j) T(r_{ij}) Y(r_{ij}) \delta(\mathbf{r}_{kj})$$
$$\left. - \frac{3m_\pi^2}{4\pi} \boldsymbol{\sigma}_k \cdot \hat{\mathbf{r}}_{kj}\, \hat{\mathbf{r}}_{kj} \cdot (\boldsymbol{\sigma}_j \times \boldsymbol{\sigma}_i) T(r_{kj}) Y(r_{kj}) \delta(\mathbf{r}_{ij}) + \boldsymbol{\sigma}_i \cdot (\boldsymbol{\sigma}_k \times \boldsymbol{\sigma}_j) \delta(\mathbf{r}_{ij}) \delta(\mathbf{r}_{kj}) \right] . \tag{5}$$

It's worth mentioning that the two-pion exchange V_C, which could be expected to be long-range from the Feynman diagram in Fig. 1, clearly also contains terms that are short-range and intermediate-range.

As in all nuclear-force formulations, the local 3N forces need to be regulated. Similarly to what is done in the NN sector, we regularize the 3N forces by smearing out the δ functions:

$$\delta(\mathbf{r}) \;\rightarrow\; \delta_{R_{3\mathrm{N}}}(\mathbf{r}) = \frac{1}{\pi\Gamma(3/4)R_{3\mathrm{N}}^3} e^{-(r/R_{3\mathrm{N}})^4} , \tag{6}$$

where $R_{3\mathrm{N}}$ is the 3N cutoff. Also, the long-range contributions are multiplied with a long-range regulator:

$$Y(r) \;\rightarrow\; Y(r)\left(1 - e^{-(r/R_{3\mathrm{N}})^4}\right) . \tag{7}$$

The NN and 3N cutoffs can be and have been varied: this reflects the fundamental fact that there is not one true nuclear potential: cutoffs are arbitrary, so it is worth probing what effect, if any, this technical feature has on the physics. The NN cutoff was varied between $R_0 = 1.0 - 1.2\,\mathrm{fm}$ [10, 13]. Smaller cutoffs lead to the appearance of spurious bound states, while larger cutoffs cut off the (physically significant) pion exchanges.

2.2. Results

Neutron matter was chosen as the first system to study using the new local chiral 3N forces because in it the structure of the 3N interaction is considerably simplified. Specifically, for $R_{3\mathrm{N}} \rightarrow 0$ the short-range and intermediate-range parts of V_C (as well as the V_E and V_D

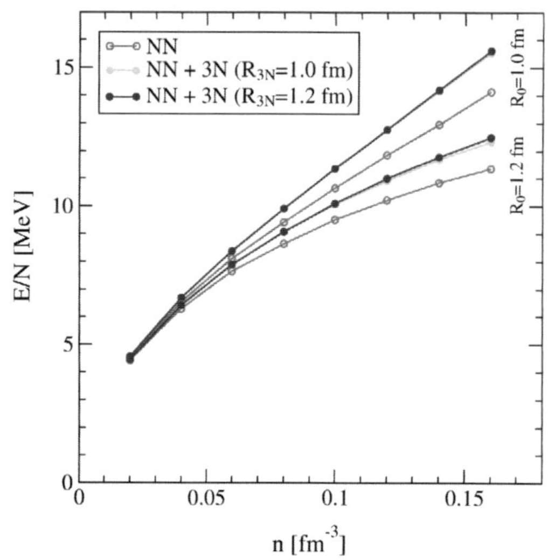

Figure 2. Energy per particle vs density for neutron matter at N^2LO, including NN forces and the 3N interaction V_C . The NN and 3N cutoffs are varied in the range $1.0 - 1.2$ fm.

contributions) vanish in neutron matter. Thus, their contribution in neutron matter for finite cutoffs is only a regulator effect. Consistently with what was done for the NN cutoff $R_0 = 1.0 - 1.2$ fm, we have varied the 3N cutoff in the range $R_{3N} = 1.0 - 1.2$ fm.

In Fig. 2 we show AFDMC results from Ref. [29] for the equation of state of neutron matter using chiral NN and 3N forces at N^2LO. Specifically, the energy per particle is shown as a function of density including NN forces and the 3N V_C interaction. The NN and 3N cutoffs have been varied in the same range, $1.0 - 1.2$ fm. For the softer NN potential ($R_0 = 1.2$ fm, lower lines) the energy per particle is $12.3 - 12.5$ MeV at saturation density for the different 3N cutoffs. The corresponding NN-only energy is 11.4 MeV, so the 3N V_C has an impact of ≈ 1 MeV. For the harder NN potential ($R_0 = 1.0$ fm, upper lines) the energy per particle is $15.5 - 15.6$ MeV (cf. 14.1 MeV for an NN-only calculation), so the effect the 3N V_C is ≈ 1.5 MeV. The variation of the total energy with the 3N cutoff is ≈ 0.2 MeV, much smaller than the variation due to the NN cutoff.

We note that the magnitude of the local 3N two-pion-exchange V_C contribution (no more than 1.5 MeV at saturation density), is smaller than the momentum-space/nonlocal-regulator contribution, which is typically $\sim$ 4 MeV [8]. This difference between 1.5 and 4 MeV is presumably due to the locality vs non-locality of the regulators in the two approaches. Similar conclusions have been drawn using coupled-cluster theory [12].

3. Static response

3.1. Formalism

Quantum Monte Carlo studies of neutron matter have also recently tackled a situation that probes further aspects of the interaction, namely the static-response problem [30]. The static response of neutron matter is synonymous to the behavior of an extended neutron system in the presence of a periodic external potential. This is equivalent to a neutron-star crust: there, unbound neutrons interact strongly with each other and with a lattice of nuclei. The unbound neutrons experience the nuclei as a periodic modulation, which one can model as a one-body potential. Approaches to this general problem were until now limited only to mean-field approaches [31, 32]. Note also that the static response of neutron matter is closely analogous to that of the three-dimensional electron gas and to liquid ^{4}He at vanishing pressure and temperature [33]. Similarly, cold fermionic atomic gases in optical lattices provide another experimental realization of related physics [34, 35].

We start from the following microscopic Hamiltonian:

$$H = -\frac{\hbar^2}{2m}\sum_i \nabla_i^2 + \sum_{i<j} V_{ij} + \sum_{i<j<k} V_{ijk} + V_{\text{ext}}\,. \tag{8}$$

In addition to a non-relativistic kinetic energy, this contains a two-nucleon interaction V_{ij} and a three-nucleon interaction V_{ijk} (Argonne v8' and Urbana IX, respectively). The last term corresponds to the external periodic potential: $V_{\text{ext}} = \sum_i V_i$, where $V_i = 2v_{\mathbf{q}}\cos(\mathbf{q}\cdot\mathbf{r}_i)$. The situation can be viewed as the application of a static (i.e., not dynamic) external potential to an unperturbed Hamiltonian. The periodicity of the potential is contained in q. Continuum QMC simulations for extended systems are carried out inside a simulation volume with periodic boundary conditions. It is important to ensure that the box length is an integer multiple of the period of the external potential.

We now discuss the fundamentals of response theory. One typically starts from an unperturbed system characterized by a Hamiltonian $\hat{H}_0$ (here, the first three terms in Eq. (8)). The last term in Eq. (8) then represents the coupling of an external potential $V_{\text{ext}}(\mathbf{r})$ to the one-body density operator $\hat{n} = \sum_{i=1}^N \delta(\mathbf{r}-\mathbf{r}_i)$. It is easy to see that $\int d^3r\, \hat{n}(\mathbf{r})V_{\text{ext}}(\mathbf{r})$ is merely a rewriting of the last term in Eq. (8). One can define the linear density-density static response function using the functional expansion of the perturbed density $n_{\text{tot}}(\mathbf{r})$ in terms of $V_{\text{ext}}(\mathbf{r})$:

$$n_{\text{tot}}(\mathbf{r}) = n_0 + \int d^3r'\; \chi(\mathbf{r}'-\mathbf{r})V_{\text{ext}}(\mathbf{r}')\,. \tag{9}$$

This $\chi(\mathbf{r}'-\mathbf{r})$ is the density-density response function of order 1, so a $\chi^{(2)}$ response function would involve two $V_{\text{ext}}(\mathbf{r})$ terms, and so on. It is possible to write down a corresponding expansion for the perturbed energy E_{tot}. This can then be re-expressed in terms of the Fourier components $v_{\mathbf{q}}$ of the modulating potential, $V_{\text{ext}}(\mathbf{r}) = \sum_{\mathbf{q}} v_{\mathbf{q}}\exp(i\mathbf{q}\cdot\mathbf{r})$, and the Fourier transform of the response, $\chi(q)$, to give:

$$E_{\text{tot}} = E_0 + \frac{\chi(q)}{n_0}v_{\mathbf{q}}^2 + C_4 v_{\mathbf{q}}^4 + \cdots\,. \tag{10}$$

Note that this expression explicitly includes higher-order response functions (hidden in the higher-order coefficients). The response function $\chi(q)$ is related to an energy integral over the dynamic structure factor, familiar from standard response theory.

3.2. Results

One can carry out AFDMC simulations for a fixed value of q: these would vary the one-body strength $2v_q$. For that specific q, then, Eq. (10) can be used to extract the $\chi(q)$ by fitting in powers of v_q. Then, changing q this process can be repeated, leading to a better understanding of the behavior of $\chi(q)$. This is clearly a computationally intensive approach, requiring dozens of simulations of ~ 100 strongly interacting particles. The results from Ref. [30] are shown in Fig. 3. The q_F used is the Fermi wave vector: the values of q/q_F available to us are prescribed by the need to respect translational invariance.

In order to provide a physical interpretation for these QMC results, Fig. 3 also shows the static response function of a non-interacting Fermi gas, known as the Lindhard function:

$$\chi_L = -\frac{mq_F}{2\pi^2\hbar^2}\left[1 + \frac{q_F}{q}\left(1-\left(\frac{q}{2q_F}\right)^2\right)\ln\left|\frac{q+2q_F}{q-2q_F}\right|\right]\,. \tag{11}$$

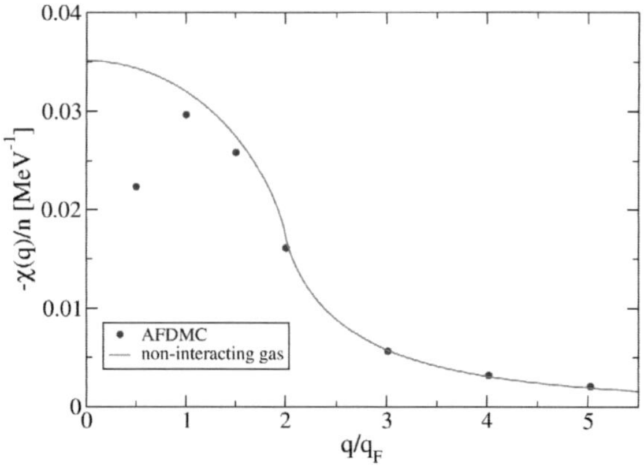

Figure 3. Static density-density linear response function of neutron matter. Points follow from AFDMC simulations using NN and 3N interactions, as well as several one-body strengths and periodicities. The line is the Lindhard function, which corresponds to the response of a non-interacting Fermi gas.

Some general features are immediately obvious. First, at large q (i.e., for the case of several clusters) the AFDMC results match the Lindhard values very well. This reflects the fact that in that scenario the interactions are vastly outweighed by the effect of the confining potential. Second, at low q we see a clear suppression, bringing to mind the case of the three-dimensional electron gas (3DEG). However, it's important to note that in the 3DEG the response function goes to 0 at $q = 0$. It is easy to see, e.g. using the random-phase approximation, that this result is specific to the Coulomb interaction, i.e., a finite-range potential changes this trend. Qualitatively, as can be seen from Eq. (10), a suppression with respect to the non-interacting gas means that the modulated interacting gas highlights the repulsion between the particles (when they are brought closer together by the external potential). It is important to note that the problem being solved here is always strongly coupled: even when $q \to 0$ the one-body strengths used extend to large fractions of the Fermi energy.

4. Summary & Conclusions

In this contribution we have touched upon two major recent developments in the study of neutron matter using QMC methods: i) local chiral forces, and ii) the static-response problem. Starting with the former, we briefly discussed the motivation behind chiral EFT and its fusing with QMC methods. We then gave the expressions describing local chiral 3N forces. After that, we turned to results for the ground-state energy of pure infinite neutron matter. We varied the 3N cutoff from $R_{3\mathrm{N}} = 1.0\,\mathrm{fm}$ to $R_{3\mathrm{N}} = 1.2\,\mathrm{fm}$ and found a much smaller dependence on this cutoff than on the NN cutoff, R_0. This finding, along with similar explorations of regulator effects in neutron matter, has motivated further studies of specific regulator choices. We then turned to the second recent advance, namely the study of periodically modulated strongly interacting neutron matter. After a brief overview of the basics of response theory, we showed the first QMC results on the static density-density linear reponse function of neutron matter. This required heavy-duty computations at several strengths and periodicities of the external potential. We found a suppression with respect to the response of the non-interacting gas, of a form that is special to the nuclear problem. This has opened up several new possibilities in the study of the interplay between matter and nuclei in neutron-star crusts.

Acknowledgments

A. G. would like to thank the Program Committee and the Local Organizing Committee for the invitation and for their efforts toward making the XVIII International Conference on Recent Progress in Many-Body Theories a success. The work discussed here was supported in part by the Natural Sciences and Engineering Research Council (NSERC) of Canada, the Canada Foundation for Innovation (CFI), the US Department of Energy, Office of Nuclear Physics, under Contract DE-AC52-06NA25396, the NUCLEI SciDAC program, the LANL LDRD program, and the ERC Grant No. 307986 STRONGINT. Computational resources were provided by SHARCNET, NERSC, Los Alamos Open Supercomputing, and the Jülich Supercomputing Center.

References

[1] Gandolfi S, Gezerlis A and Carlson J 2015 *Ann. Rev. Nucl. Part. Sci.* **65**, 303
[2] Hebeler K, Holt J D, Menendez J and Schwenk A 2015 *Ann. Rev. Nucl. Part. Sci.* **65**, 457
[3] Carlson J, Morales J Jr, Pandharipande V R and Ravenhall D G 2003 *Phys. Rev. C* **68**, 025802
[4] Gandolfi S, Illarionov A Yu, Schmidt K E, Pederiva F and Fantoni S 2009 *Phys. Rev. C* **79**, 054005
[5] Gezerlis A and Carlson J 2010 *Phys. Rev. C* **81**, 025803
[6] Gandolfi S, Carlson J and Reddy S 2012 *Phys. Rev. C* **85**, 032801
[7] Baldo M, Polls A, Rios A, Schulze H J and Vidaña I 2012 *Phys. Rev. C* **86**, 064001
[8] Hebeler K and Schwenk A 2010 *Phys. Rev. C* **82**, 014314
[9] Tews I, Krüger T, Hebeler K and Schwenk A 2013 *Phys. Rev. Lett.* **110**, 032504
[10] Gezerlis A, Tews I, Epelbaum E, Gandolfi S, Hebeler K, Nogga A and Schwenk A 2013 *Phys. Rev. Lett.* **111**, 032501
[11] Coraggio L, Holt J W, Itaco N, Machleidt R and Sammarruca F 2013 *Phys. Rev. C* **87**, 014322
[12] Hagen G, Papenbrock T, Ekström A, Wendt K A, Baardsen G, Gandolfi S, Hjorth-Jensen M and Horowitz C J 2014 *Phys. Rev. C* **89**, 014319
[13] Gezerlis A, Tews I, Epelbaum E, Freunek M, Gandolfi S, Hebeler K, Nogga A and Schwenk A 2014 *Phys. Rev. C* **90**, 054323
[14] Carbone A, Rios A and Polls A 2014 *Phys. Rev. C* **90**, 054322
[15] Lovato A, Benhar O, Fantoni S and Schmidt K E 2012 *Phys. Rev. C* **85**, 024003
[16] Pudliner B S, Pandharipande V R, Carlson J, Pieper S C and Wiringa R B 1997 *Phys. Rev. C* **56**, 1720
[17] Pieper S C and Wiringa R B 2001 *Annu. Rev. Nucl. Part. Sci.* **51**, 53
[18] Nollett K M, Pieper S C, Wiringa R B, Carlson J and Hale G M 2007 *Phys. Rev. Lett.* **99**, 022505
[19] Schmidt K E and Fantoni S 1999 *Phys. Lett. B* **446**, 99
[20] Epelbaum E, Hammer H W and Meißner U G 2009 *Rev. Mod. Phys.* **81**, 1773
[21] Entem D R and Machleidt R 2011 *Phys. Rept.* **503**, 1
[22] Hammer H W, Nogga A, and Schwenk A 2013 *Rev. Mod. Phys.* **85**, 197
[23] van Kolck U 1994 *Phys. Rev. C* **49**, 2932
[24] Epelbaum E, Nogga A, Glöckle W, Kamada H, Meißner U G and Witała H 2002 *Phys. Rev. C* **66**, 064001
[25] Epelbaum E, Krebs H, Lee D and Meißner U G 2009 *Eur. Phys. J. A* **40**, 199
[26] Roggero A, Mukherjee A and Pederiva F 2014 *Phys. Rev. Lett.* **112**, 221103
[27] Wlazłowski G, Holt J W, Moroz S, Bulgac A and Roche K J 2014 *Phys. Rev. Lett.* **113**, 182503
[28] Lynn J, Carlson J, Epelbaum E, Gandolfi S, Gezerlis A and Schwenk A 2014 *Phys. Rev. Lett.* **113**, 192501
[29] Tews I, Gandolfi S, Gezerlis A and Schwenk A 2015 Quantum Monte Carlo calculations of neutron matter with chiral three-body forces preprint arXiv:1507.05561
[30] Buraczynski M and Gezerlis A 2015 Static Response of Neutron Matter preprint arXiv:1510.06417
[31] Chamel N 2011 *Phys. Rev. C* **85**, 035801
[32] Chamel N, Page D and Reddy S 2013 *Phys. Rev. C* **87**, 035803
[33] Moroni S, Ceperley D M and Senatore G 1992 *Phys. Rev. Lett.* **69**, 1837
[34] Pilati S, Zintchenko I and Troyer M 2014 *Phys. Rev. Lett.* **112**, 015301
[35] Carlson J and Gandolfi S 2014 *Phys. Rev. A* **90**, 011601(R)

Journal of Physics: Conference Series **702** (2016) 012015 doi:10.1088/1742-6596/702/1/012015

Comparative study of the *LOCV* and the *FHNC* approaches for the *nucleonic* matter problem

Azar Tafrihi [1] **and Majid Modarres** [2]

1 Department of Physics, Isfahan University of Technology, Isfahan 84156-83111, Iran
2 Physics Department, University of Tehran, North-Kargar Ave., 1439955961 Tehran, Iran

E-mail: tafrihi@cc.iut.ac.ir

Abstract. The *nucleonic* matter problem is investigated by comparing the lowest order constrained variational (*LOCV*) method with the Fermi *hypernetted* chain (*FHNC*) theory, emphasizing the role of the *LOCV* correlation functions. In this way, the central correlation functions are used in the *LOCV* formalism, for the *Bethe* homework problem. It is shown that the *LOCV* computations reasonably agree with those of *FHNC*. Moreover, the *FHNC* calculations are performed with the *LOCV* correlation functions. It is found that, assuming the *LOCV* or the parametrized correlation functions, the *FHNC* computations do not change significantly. So, one may conclude that the mentioned consistencies refer to the choice of the *LOCV* correlation functions. Because, the contribution of the many-body cluster terms can be ignored, if the *LOCV* correlation functions satisfy the normalization constraint. Then, using the $AV18$ interaction, the operator-dependent (*OD*) correlation functions are employed in the *LOCV* calculations. Note that the *LOCV OD* correlation functions are obtained by averaging over the states. It turns out that the overall behaviour of the *LOCV OD* correlation functions are similar to those of *FHNC*. Although, due to the many-body effects which are considered in the *FHNC* calculations, the *LOCV* results fairly differ from those of *FHNC*. Finally, it is worth mentioning that, unlike the recent *FHNC* calculations, the spin-orbit-dependent correlation functions are included in the *LOCV* approach.

1. Introduction

The lowest order constrained variational (*LOCV*) approach has been hitherto applied to the *nucleonic* matter problem, at zero and finite temperatures [1–19], with different two-body potentials, e.g. the Reid [20] and the *Argonne* interactions [21–25]. In these calculations [1–19], the energy of the *nculeonic* matter is obtained by truncating the cluster expansion of the energy [26, 27], at the two-body cluster term. Since, the *LOCV* formalism is fully state-dependent, in each channel, the two-body correlation functions are computed by functionally minimizing the energy, subject to the normalization constraint [28]. As it was shown in our previous papers [8, 14, 17], by imposing the normalization constraint, the contribution of the higher cluster terms than the two-body one becomes ignorable. So, it was illustrated that the *LOCV* results [1–19] fairly agree with the corresponding predictions of the sophisticated methods, i.e. the Fermi *hypernetted* chain (*FHNC*) and the Monte Carlo (*MC*) techniques.

In recent years, we compared the *LOCV* method with the *FHNC* theory, for different nuclear many-body problems [17–19]. These comparisons were based on the role of the correlation functions. At the first step, these two methods were compared against the *Fermionic* prescription of the *Bethe* homework problem [17]. The *LOCV* and the extended *LOCV*

($ELOCV$) central correlation functions were included in the $FHNC$ and the *Fermi* chain (FC) formalisms instead of the parametrized *Krotscheck* and *Takahashi* [29] (KT) ones. It should be noted that in the $ELOCV$ computations, the cluster expansion of the energy is truncated at the three-body level. So, we showed that including the $(E)LOCV$ or the KT correlation functions, the $FHNC(FC)$ predictions [30,31] do not change considerably. Moreover, it was found that the approximated $(E)LOCV$ results reasonably agree with those $FHNC$ [30,31]. It was concluded that these consistencies are originated from the $(E)LOCV$ correlation functions, which optimally satisfy the normalization constraint. In the next step, we reformulated the $LOCV$ method for a fixed number of particles in a periodic box ($PBLOCV$) [18]. It was found that the $PBLOCV$ results fairly agree with those of $FHNC$ calculations in the periodic boxes ($PBFHNC$) [32,33]. This agreement refers to the choice of correlation function in the $LOCV$ method, i.e. the satisfaction of the normalization constraint in the periodic boxes. Furthermore, the finite size effects, which can be used to estimate the finite size effects in the auxiliary field diffusion Monte Carlo ($AFDMC$) simulations, were investigated in the $PBFLOCV$ approach [18]. As it was expected, by increasing the dimensions of the periodic boxes, the $PBLOCV$ results became independent of the number of particles [18]. Finally, we rewritten the channel-dependent $LOCV$ *nucleonic* matter correlation functions in terms of the first eight operators of the $AV18$ potential [19]. In this way, the $LOCV$ operator-dependent correlation functions [19] were compared with those of $FHNC$ at single-operator chain ($FHNC/SOC$) approximation [34–36]. It was demonstrated that there are some discrepancies between the $LOCV$ and the $FHNC/SOC$ correlation functions. Because, unlike the $FHNC/SOC$ correlation functions, the $LOCV$ results are defined to fulfil the two-body cluster approximation. Similarly, due to the many-body effects, the $FHNC/SOC$ *nucleonic* matter energies became different from those of $LOCV$. However, one may hope that similar to the *Bethe* homework problem [17], inserting the $LOCV$ operator-dependent correlation functions in the $FHNC/SOC$ formalism: (i) The $FHNC/SOC$ energy does not change significantly. (ii) The $FHNC/SOC$ results converge faster.

According to the above works [17–19], we illustrated that the $LOCV$ correlation functions, which satisfy the normalization constraint, have several advantages with respect to those of $FHNC(/SOC)$:

(i) The $LOCV$ two-body correlation functions are state-dependent, i.e. they are different in each nucleon-nucleon state. On the other hand, the $FHNC/SOC$ correlation functions are state-independent. Because, using the channel-dependent correlation functions, the many-body calculations become complicated. So, according to the reference [37], it is expected that the $LOCV$ channel-dependent correlation functions, which satisfy the normalization constraint, could be more optimal than those of $FHNC/SOC$.

(ii) The $FHNC/SOC$ correlation functions can not directly be obtained by the minimization of the many-body energy [34]. The $FHNC/SOC$ correlation functions are calculated at the lowest order approximation [34]. Then, using some variational parameters, the contribution of the many-body terms is included in the $FHNC/SOC$ correlation functions [34]. While, the $LOCV$ correlation functions are obtained by functionally minimizing the energy, at the two-body cluster approximation, subject to the normalization constraint. Note that, unlike the $FHNC/SOC$ technique, in the $LOCV$ formalism, the cluster expansion of the energy is truncated at the lowest order. So, the $LOCV$ energy and correlation function could be reported at the same approximation, i.e. at the two-body cluster approximation.

(iii) The $LOCV$ channel-dependent correlation function includes the spin-orbit-dependent correlation functions. While, in recent $FHNC/SOC$ calculations [38,39], the contribution of the spin-orbit-dependent correlation functions has been ignored. Because, the treatment of the spin-orbit correlation functions in the $FHNC/SOC$ formalism is associated with uncertainties and difficulties [40].

So, in this paper, we present a comparative study of the *LOCV* [17,19] and the *FHNC* approaches [30,31,35,36,39,41], by concentrating on the infinite systems, i.e. the pure neutron (symmetric nuclear) matter. It should be noted that the principal aim of this paper is to illustrate that the approximated *LOCV* approach is capable of reproducing the *FHNC(/SOC)* and the *Monte Carlo* calculations in a more simpler way. However, it is shown that, in the case of the spin-orbit-dependent terms, the *LOCV* method is more reliable than the sophisticated *FHNC/SOC* and *Monte Carlo* approaches. In this way, the plan of this paper is as follows: (i) In the Section 2, first, we give the *LOCV* formalism for the *Bethe* homework problem [17]. Then, the *LOCV* formalism for the *nucleonic* matter system, using the *AV*18 interaction, is represented [19]. Also, the *LOCV* operator-dependent correlation functions are extracted from the *LOCV* channel-dependent formalism [19]. (ii) In the Section 3, the *LOCV* results [17,19] are shown in comparison with those of *FHNC(/SOC)* [30,31,35,36,39,41] and Monte Carlo [25,39,42]. (iii) In addition, the corresponding discussions and conclusions are given in the Section 4. Note that in this section: (a) The updated versions of our previous calculations for the *Bethe* homework problem [14], are represented. (b) The *LOCV nucleonic* matter computations of the reference [16] are reviewed.

2. The *LOCV* formalisms

2.1. The LOCV formalism for the Bethe homework problem

The Hamiltonian of the *Bethe* homework problem for N neutrons can be written as follows [17]:

$$H = \sum_{i=1}^{N} \frac{p_i^2}{2m} + \frac{1}{2}\sum_{i\neq j}^{N} v(ij), \tag{1}$$

where $v(ij)$ is given as [17]:

$$v(ij) \equiv v_B(r_{ij}) = 9263.1 \frac{e^{-4.9 r_{ij}}}{r_{ij}} MeV, \tag{2}$$

and r is measured in fm. Note that the *Bethe* potential is the repulsive part of the Reid soft-core $^1S_0$ interaction [20]. An ideal Fermi gas type wave function is used in the *LOCV* (*ELOCV*) method, for the single particle states. Furthermore, using the variational techniques, the wave function of the interacting system is defined as [4–19]:

$$\psi = F\phi, \tag{3}$$

where

$$F = S\prod_{i>j} f(ij). \tag{4}$$

Note that ϕ is the *Slater* determinant of plane waves and S is a symmetrizing operator. Similar to the *Bethe* potential (equation (2)), the correlation functions $f(ij)$ are assumed to depend on the relative distance of two interacting particles.

Now, considering the cluster expansion of the energy [26,27], one can expand the expectation value of the equation (1) as follows:

$$E[f] = \frac{1}{N}\frac{\langle\psi|H|\psi\rangle}{\langle\psi|\psi\rangle} = T_F + E_{MB}, \tag{5}$$

where, T_F is the Fermi gas kinetic energy, which is defined in terms of the Fermi momentum k_f ($= (3\,\pi^2\,\rho)^{\frac{1}{3}}$), and E_{MB} is the contribution of the many-body clusters, i.e. $E_{MB} = E_2 + E_3 + ... + E_N$.

In the $(E)LOCV$ method, the cluster expansion of the energy is truncated at the two(three)-body cluster term $E_{2(3)}$, i.e. [27]

$$E_2 = \frac{1}{2N} \sum_{ij} \langle ij|V_{eff}|ij\rangle_a, \tag{6}$$

$$E_3 = [E_{3(2)}] + E_{3(3)} + E_{4(2)} = [E_{3h} + E_{3hh}] + E_{3t} + E_{4h}, \tag{7}$$

where [27],

$$V_{eff}(12) = -\frac{\hbar^2}{2m} [f(12), [\nabla_{12}^2, f(12)]] + f(12)\, v(12)\, f(12), \tag{8}$$

and E_{3h}, E_{3hh}, E_{3t} and E_{4h} are given in the equations (11) to (14) of the reference [17]. Note that $|ij>_a$ means that the neutron-neutron state is asymmetric. Applying the plane waves as the non-interacting wave functions, in the equation (6) and the equations (11) to (14) of the reference [17], one arrives at the equations (3.12) to (3.14) and (3.16) of the reference [27] for the two- and the three-body cluster energies.

As it was mentioned in the introduction, in the $LOCV$ approach, the two-body cluster energy is minimized, subject to the normalization constraint [28], i.e.

$$\xi \equiv < \psi|\psi > -1 = \rho \int (1 - \mathcal{G}_2(r_{12}))\, d\mathbf{r}_{12} - 1, \tag{9}$$

where the radial distribution function $\mathcal{G}_2(r_{12})$ is written as follows [27]:

$$\mathcal{G}_2(r_{12}) = f^2(r_{12}) \sum_{n=2}^{N} [\triangle\, \mathcal{G}_2(r_{12})]_n, \tag{10}$$

and $[\triangle\, \mathcal{G}_2(r_{12})]_{n=2,3}$ are given in the equations (20) to (21) of the reference [17]. Note that in the $(E)LOCV$ calculations, $[\triangle\, \mathcal{G}_2(r_{12})]_{n=2(2,3)}$ is (are) included in the equation (10). Moreover, assuming the normalization constraint (equation (9)), the Lagrange multiplier λ is introduced in the $LOCV$ formalism. In this way, the functional $L[f] = \{E_2 + \lambda\, \langle\psi|\psi\rangle\}[f]$ is minimized with respect to $f(r)$, by solving the Euler-Lagrange equation, i.e. [17]

$$\frac{\partial L}{\partial f(r)} - \frac{\partial}{\partial r}\frac{\partial L}{\partial f'(r)} = 0, \tag{11}$$

which leads to the following differential equation [17]:

$$g''(r) - [\frac{A''(r)}{A(r)} + \frac{m}{\hbar^2}\,(v(r) - \lambda)]\, g(r) = 0, \tag{12}$$

where

$$g(r) = A(r)\, f(r), \tag{13}$$

and

$$A^2(r) = r^2\, \mathcal{G}_F(r). \tag{14}$$

In the above equation, $\mathcal{G}_F(r)(= 1 - \frac{1}{2} l^2(k_f r))$ is the Fermi gas radial distribution function which is defined in terms of the spherical Bessel functions, i.e. $l(x) = \frac{3\, J_1(x)}{x}$. It should be stated that regarding the $ELOCV$ normalization constraint, the Euler-Lagrange equation (12) becomes an integro-differential equation in terms of $f(r)$.

The Euler-Lagrange differential equation (12) is solved while $(ln(f(r))' = (ln(f_p(r))'$. Note that $f_p(r)(= \mathcal{G}_F^{-\frac{1}{2}}(r))$ is the *Pauli* function. Then, for $r > d$ (d is the healing distance), $f(r) = f_p(r)$. In a similar way, the *ELOCV* integro-differential equation can iteratively be solved until the equation (10) is satisfied for $n = 3$. Then, the contribution of E_3 is added to that of E_2. It should be noticed that E_3 is an estimate to the three-body cluster energy, i.e. E_2+E_3 can not be considered as the upper bound to the true energy. However, if the contribution of E_3 becomes very small in comparison with that of E_2, then it can be concluded that $T_F + E_2$ is a good approximation to the upper bound ground state energy of *Bethe* homework problem.

2.2. The LOCV formalism for the nucleonic matter with the AV18 potential
The *LOCV* formalism, for the *nucleonic* matter system, using the *AV*18 potential, is similar to the formalism of the previous subsection. Although, one should consider the following points: (i) In the equation (2), the *AV*18 interaction [22] is considered instead of the *Bethe* potential. (ii) In the equation (4), $f(ij)$ is assumed to be state-dependent. So, the *LOCV nucleonic* matter two-body correlation function is defined as follows [19]:

$$F(12) = \sum_{JSTT_z} |JSTT_z > f_{JSTT_z}(12) < JSTT_z|, \tag{15}$$

where J, S and $T(T_z)$ are the angular momentum, the spin and the (z axis projection of) *isospin* of the pair of nucleons. Note that $T, T_z = 1, -1$, in the pure neutron matter computations. Moreover, in the decoupled (DC) channels, i.e. for $S = 0$ and $S = 1$, $J = 0$ (L) states, $f_{JSTT_z}(12)$ is central, i.e. $f^{DC}_{JSTT_z}(12) = f^{(1)}_{JSTT_z}(r_{12})$. On the other hand, in the coupled (CP) channels $^3S_1 -^3 D_1$ and $^3P_2 -^3 F_2$, $f_{JSTT_z}(12)$ are respectively given as follows [19]:

$$f^{CP}_{1100}(12) = f^{(21)}_{1100}(r_{12})Q + f^{(31)}_{1100}(r_{12})\overline{Q}, \tag{16}$$

$$f^{CP}_{211T_z}(12) = f^{(22)}_{211T_z}(r_{12})P^{L=1} + f^{(32)}_{211T_z}(r_{12})P^{L=3}, \tag{17}$$

where $Q(\equiv 1 - \overline{Q}) = \frac{2}{3} + \frac{1}{6}S_{12}$ and $P^L = |L >< L|$. Also, $f^{(ij)}_{JSTT_z}(r_{12})$ can be written in terms of the radial and the tensor(spin-orbit)-dependent correlation functions, i.e. [11, 19],

$$f^{(i1)}_{1100}(r_{12}) = f^{r}_{1100}(r_{12}) + a_1 f^{t}_{1100}(r_{12}), \tag{18}$$

$$f^{(i2)}_{211T_z}(r_{12}) = f^{r}_{211T_z}(r_{12}) + a_2 f^{b}_{211T_z}(r_{12}), \tag{19}$$

where, for $i = 2$ (3), $a_1 = 2\,(-4)$ and $a_2 = 1\,(-4)$. Note that similar to the equation (16), one can employ the tensor-dependent correlation function in the $^3P_2 -^3 F_2$ channel.

With regard to the equation (15) and the two-body potential *AV*18, the two-body cluster energy, i.e. the equation (6), can be written as follows [19]:

$$E_2 = \frac{\rho}{2}\int d^3r_{12}\{\sum_{JLSTT_zT_{z_1}T_{z_2}} \varpi v^{JLSTT_z}_{eff}(r_{12}) I_{L,T_z}(k_f r_{12})\}, \tag{20}$$

where $\varpi = \frac{k_f^2}{\pi^5\rho^2}| < \frac{1}{2}T_{z_1}\frac{1}{2}T_{z_2}|TT_z > |^2(2J+1)\frac{1}{2}(1-(-1)^{L+S+T})$ and $I_{L,T_z}(k_f r_{12})$ is defined in the reference [16]. In addition, $v^{JLSTT_z}_{eff}(r_{12})$ for the coupled and decoupled channels were given in the references [11, 16, 19].

Now, the two-body cluster energy is minimized, subject to the normalization constraint, i.e. equation (9), which for the state-dependent correlation functions, is obtained by the following relation [11, 19]:

$$\xi = \rho \int d^3 r_{12} \{ \sum_{JLSTT_z T_{z_1} T_{z_2}} \varpi (f^2_{p,T_z}(r_{12}) - f^2_{JSTT_z}(r_{12})) I_{L,T_z}(k_F r_{12}) \} - 1, \quad (21)$$

where $f_{p,T_z}(r_{12})$ is the *Pauli* function which defines the long-range behaviour of the *LOCV* correlation functions. Note that $f_{p,\pm 1}(r_{12}) = [\mathcal{G}_F(r_{12})]^{-\frac{1}{2}}$ and $f_{p,0}(r_{12}) = 1$. So, minimizing the equation (20), subject to the equation (21), one arrives at the following Euler-Lagrange equations for the decoupled, the $^3P_2 - ^3F_2$ and the $^3S_1 - ^3D_1$ channels, respectively [11, 16, 19]:

$$\mathcal{G}^{(1)''}_{JSTT_z}(r_{12}) - \mathcal{G}^{(1)}_{JSTT_z}(r_{12}) \{ \frac{a^{(1)''}_{JSTT_z}(r_{12})}{a^{(1)}_{JSTT_z}(r_{12})} + \frac{m}{\hbar^2}(v^{(1)}_{JSTT_z}(r_{12}) + \lambda) \} = 0, \quad (22)$$

$$\mathcal{G}^{(i2)''}_{211T_z}(r_{12}) - \mathcal{G}^{(i2)}_{211T_z}(r_{12}) \{ \frac{a^{(i2)''}_{211T_z}(r_{12})}{a^{(i2)}_{211T_z}(r_{12})} + \frac{m}{\hbar^2}(v^{(i2)}_{211T_z}(r_{12}) + \lambda) \} = 0, \quad (23)$$

$$\mathcal{G}^{(i1)''}_{1100}(r_{12}) - \mathcal{G}^{(i1)}_{1100}(r_{12}) \{ \frac{a^{(i1)''}_{1100}(r_{12})}{a^{(i1)}_{1100}(r_{12})} + \frac{m}{\hbar^2}(v^{(i1)}_{1100}(r_{12}) + \lambda) + \Gamma^{(i1)}_{1100}(r_{12}) \} + \Delta^{(i1)}_{1100} \mathcal{G}^{(j1)}_{1100}(r_{12}) = 0, \quad (24)$$

where for $i = 2, 3$, $j = 3, 2$ and

$$\mathcal{G}^{(1,i1,i2)}_{JSTT_z}(r_{12}) = a^{(1,i1,i2)}_{JSTT_z}(r_{12}) f^{(1,i1,i2)}_{JSTT_z}(r_{12}), \quad (25)$$

$$a^{(1,i2)}_{JSTT_z}(r_{12}) = (k_F r_{12})^2 I_{L,T_z}(k_F r_{12}), \quad (26)$$

$$a^{(21(31))}_{1100}(r_{12}) = (k_F r_{12})^2 \{ \beta(\gamma) I_{J-1,T_z}(k_F r_{12}) + \gamma(\beta) I_{J+1,T_z}(k_F r_{12}) \}. \quad (27)$$

Note that $v^{(i2)}_{JSTT_z}(r_{12})$, β and γ were given in the references [11,19]. Also, $v^{(1,i1)}_{JSTT_z}(r_{12})$, $\Gamma^{(i1)}_{1100}(r_{12})$ and $\Delta^{(i1)}_{1100}$ are written in terms of $\tilde{V}$, a_α, b_α and d_α which were defined in the reference [11, 16] (see the equations (23) to (25) of the reference [16]).

Similar to the previous section, for $r \leq d$, the above Euler-Lagrange equations are iteratively solved by changing the parameter λ. Then, for $r > d$, the *LOCV* correlation functions are defined by the *Pauli* correlation functions. In this way, the two-body correlation function $F(12)$ in different channels, i.e. $f_{JSTT_z}(r_{12})$, are computed in the *LOCV* framework.

Now, one can rewrite $F(12)$ (equation (15)) in terms of the first eight operators of the $AV18$ interaction as follows:

$$F(12) = \sum_p f_p(r_{12}) O_p(12), \quad (28)$$

$$= f_c(r_{12}) + f_s(r_{12}) \sigma_1.\sigma_2 + f_t(r_{12}) \tau_1.\tau_2 + f_{\sigma\tau}(r_{12}) \sigma_1.\sigma_2 \tau_1.\tau_2 + f_T(r_{12}) S_{12} + \quad (29)$$
$$f_{Tt}(r_{12}) S_{12} \tau_1.\tau_2 + f_b(r_{12}) L.S + f_{bt}(r_{12}) L.S \tau_1.\tau_2,$$

where, regarding the equations $(B1)$ to $(B6)$ of the reference [19], $f_p(r_{12})$ could be obtained by evaluating $F(12)$ in $ST = 00, 01, 10, 11$. Since the *LOCV* correlation functions are $JSTT_z$-dependent, we find $F_{ST}(12)$ by the use of the following averaging expression [19]:

$$\bar{F}_{ST}(12) = \frac{\sum_{JLT_z T_{z_1} T_{z_2}} \varpi f_{JSTT_z}(r_{12}) I_{L,T_z}(k_F r_{12})}{\sum_{JLT_z T_{z_1} T_{z_2}} \varpi I_{L,T_z}(k_F r_{12})}. \quad (30)$$

So, considering the equations $(B1)$ to $(B10)$ of the reference [19] and the above relation, one can extract the operator-dependent correlation functions $f_p(r_{12})$ (equation (28)) from the *LOCV* channel-dependent formalism.

3. Results, discussions and conclusions

3.1. The Bethe homework problem

In this subsection, we present the $LOCV$ results for the *Bethe* homework problem in comparison with those of $FHNC$ [30, 31]. To this respect, first, we present the $(E)LOCV$ correlation functions in comparison with the corresponding parametrized correlation function KT, i.e. $f_{KT}(r) = exp(-\frac{b}{2\,r}exp(-\frac{r}{b}))$, which is used in the $FHNC$ calculations of the references [30,31]. Note that parameter b at different densities were given in the Table 1 of the reference [17]. Then, we compare the $(E)LOCV$ equation of state for the *Bethe* homework problem with the corresponding results of $FHNC$ [30,31].

In the figure 1, the $(E)LOCV$ correlation function at $\rho = 0.6\,fm^{-3}$ are compared with that of KT. To show the effects of the normalization constraint (equation (9)), we impose two different constraints: 1. $\xi(\equiv <\psi|\psi> -1) = 0$. This means that in the cluster expansion of $<\psi|\psi>\,(= 1 + \chi_2 + \chi_3 + ...)$ [8], the two-body (two- and three-body) cluster term(s), i.e. $\chi_2\,(\chi_2+\chi_3)$, is (are) considered to become zero in the $(E)LOCV$ calculations. The corresponding calculations are labelled by $(E)LOCV$. 2. $\xi(\equiv <\psi|\psi> -1) = 1$, i.e. in the cluster expansion of $<\psi|\psi>$, $\chi_2\,(\chi_2+\chi_3)$, is (are) considered to become 1 in the $(E)LOCV$ calculations. However, assuming $\chi_2 + \chi_3 = 1$, the $ELOCV$ formalism does not work. The $LOCV$ computations, applying $\chi_2 = 1$, are labelled by $LOCV(\chi_2 = 1)$. Note that this figure is an updated version of the Figure 2 of the reference [17].

As it is illustrated in the figure 1, the $(E)LOCV$ correlation functions for $\chi_2\,(\chi_2+\chi_3) = 0$ are short-range in comparison with that of KT, which is employed in the $FHNC$ computations. On the other hand, similar to the KT correlation function, considering $\chi_2 = 1$, the corresponding $LOCV$ result, i.e. $LOCV(\chi_2 = 1)$, becomes long-range. So, it can be concluded that assuming the KT correlation function, the contribution of the many-body cluster terms in the cluster expansion of $<\psi|\psi>$ becomes large, i.e. the KT correlation function can not satisfy the normalization constraint (equation (9)). It is obvious that the KT correlation function should be long-range to be able to be used in the many-body calculations. Also, we know that the mentioned function is defined in terms of the parameter b, which is fixed by the minimization of the many-body energy. But, unlike the $LOCV$ case, the mentioned minimization has not been done, subjected to any condition, e.g. the normalization constraint. Moreover, as it was mentioned in the previous section, the long-range part of the $(E)LOCV$ correlation function is defined by the *Pauli* function. So, this boundary condition makes the $(E)LOCV$ correlation functions overshoot to one. This overshoot: (i) maximizes the effect of the nucleon-nucleon interaction in the short inter-particle distances. (ii) makes the two-body distribution function $g(r)$ become exactly 1, for $r > d$ (d is the healing distance).

In the figure 2, the $(E)LOCV$ equation of state for the *Bethe* homework problem, imposing $\chi_2\,(\chi_2+\chi_3) = 0$, are represented in comparison with the $FHNC$ computations of *Zabolitzky* [31] and the $LOCV$ predictions of *Owen*, *Bishop* and *Irvine* [7] (OBI). It should be mentioned that in the $FHNC$ calculations of *Zabolitzky*, the KT correlation function is employed. To clarify the significance of the normalization constraint role, the $LOCV$ calculations, assuming $\chi_2 = 1$ ($LOCV(\chi_2 = 1)$), are also shown in this figure. As it was mentioned above, the $ELOCV$ formalism does not work when $\chi_2 + \chi_3 = 1$. Moreover, the $FHNC$ computations, using the $(E)LOCV$ and the $LOCV(\chi_2 = 1)$ correlation functions, which are respectively labelled by $(E)FHNC$ and $FHNC(\chi_2 = 1)$, are plotted for comparison. Note that: (i) The equations (28) to (31) of the reference [17], are used for the $FHNC$ calculations with the $(E)LOCV$ correlation functions. (ii) This figure should be considered as the updated version of the Figure 3 of the reference [17]. (iii) For comparison with the available $FHNC$ and the Monte Carlo results, our computations are given at high densities, i.e. at densities larger than 6 times nuclear saturation density. (iv) The Monte Carlo calculations of *Ceperley*, *Chester* and *Kalos* (CCK) [42], which

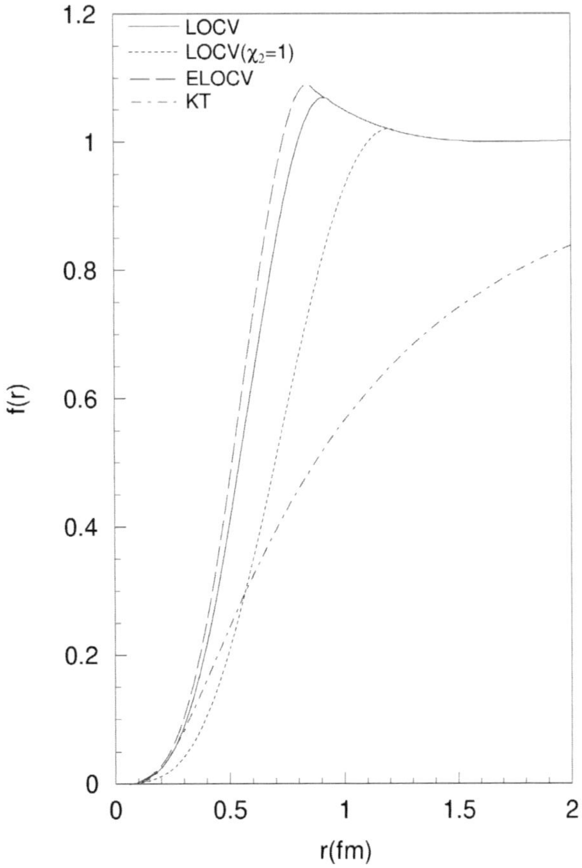

Figure 1. The $(E)LOCV$ $Bethe$ homework problem correlation function, for $\chi_2(\chi_2+\chi_3)=0$, versus those of KT and $LOCV(\chi_2=1)$. This figure should be considered as the updated version of the Figure 2 of the reference [17].

should be considered as the exact results, are shown by dotted line in this figure.

Referring to the figure 2, it is apparent that the $LOCV$ results are expectedly similar to those of OBI. Note that OBI also used the $LOCV$ formalism. Furthermore, considering $\chi_2(\chi_2+\chi_3)=0$, the $(E)LOCV$ results fairly agree with those of *Zabolitzky*. Whereas, unlike the $FHNC$ computations of *Zabolitzky*, in the $(E)LOCV$ calculations, the cluster expansion of energy is truncated, at the two(three)-body cluster term. This consistency refers to the normalization constraint, which is optimally satisfied at the two(three)-body cluster approximation. So, in the $(E)LOCV$ approach, the contribution of higher cluster terms becomes negligible. In addition, the $FHNC$ calculations with the $(E)LOCV$ correlation function, for $\chi_2(\chi_2+\chi_3)=0$, fairly agree with the corresponding computations of *Zabolitzky*. While, the former calculations are obtained with one decimal place accuracy, by the use of a simple mixing iteration method. On the other hand, the $LOCV$ results, assuming $\chi_2=1$, lie far below of the other results. Similarly, the $FHNC$ computations with the $LOCV(\chi_2=1)$ correlation function significantly differ from the corresponding results, for which the $LOCV(\chi_2=0)$ correlation

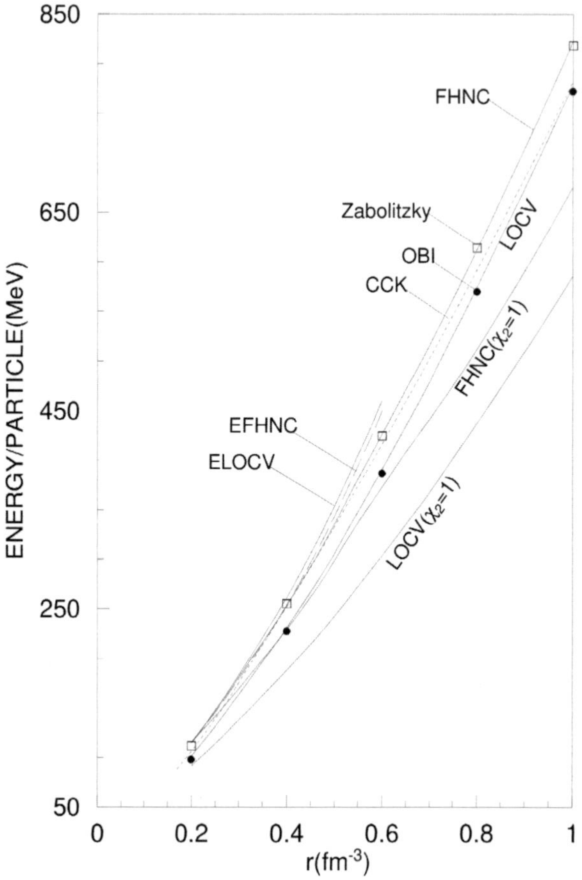

Figure 2. The $LOCV$, $ELOCV$, $FHNC$ and $EFHNC$ results in comparison with those of $LOCV(\chi_2 = 1)$ and $FHNC(\chi_2 = 1)$. For a better comparison the OBI [7], *Zabolitzky* [31] and CCK [42] calculations are also shown by dots, squares and dotted line, respectively. This figure is an updated version of the Figure 3 of the reference [17].

function is included. So, one may realize that the consistency among different approaches disappears when $\chi_2 = 1$, even if one uses the higher approximation method, i.e. the $FHNC$ approach. In the other word, the satisfaction of the normalization constraint, at the two(three)-body cluster term, leads to the reasonable agreements between the $(E)LOCV$ and the $FHNC$ results. Finally, it is worth mentioning that our approximated $(E)LOCV$ results fairly agree with those of CCK [42], especially at (low) high densities. So, one may realize that the approximated $LOCV$ method is able to achieve reasonable results for the *Bethe* homework problem in a more simpler way than the $FHNC$ and the Monte Carlo approaches.

3.2. The nucleonic matter problem for the $AV18$ interaction

At the beginning, we compare the $LOCV$ *nucleonic* matter equation of state [19] with those of $FHNC/SOC$ [39, 41] and $AFDMC$ [25, 39]. Then, the $LOCV$ operator-dependent correlation functions [19] are compared with those of $FHNC/SOC$ [35, 36].

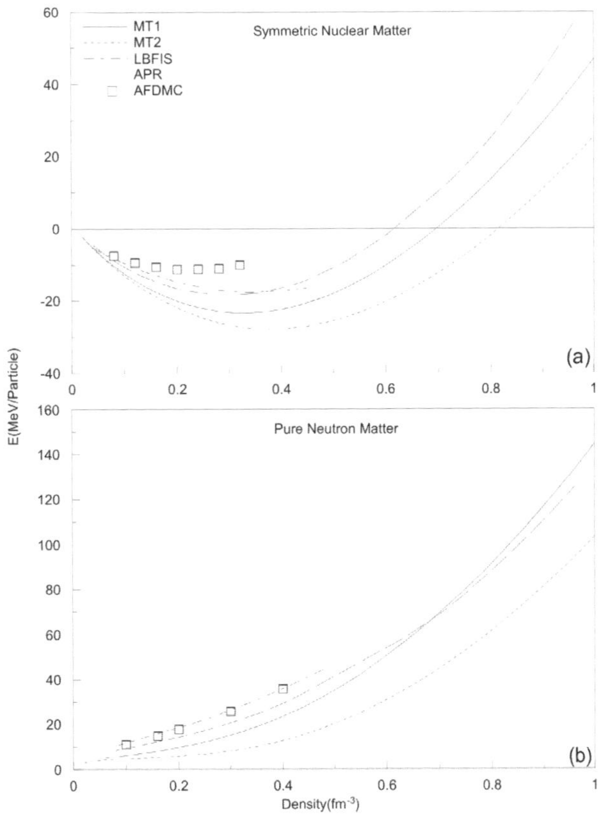

Figure 3. (a) The $LOCV$ nuclear matter equation of state [19], using the tensor(spin-orbit)-dependent correlation function in the $^3P_2 -^3 F_2$ channel, is labelled by $MT1(2)$. The corresponding $FHNC/SOC$ ($APR(LBFIS)$) [41] ([39]) and the $AFDMC$ [25] results are presented for comparison. Note that the $MT1, 2$ and the APR data are obtained, employing the $AV18$ interaction [22]. While, the $LBFIS$ and the $AFDMC$ predictions are achieved, using the $AV'7, 8$ interactions [23,25]. (b) The $LOCV$ neutron matter equations of state [19] ($MT1, 2$), for the $AV18$ [22] interaction, in comparison with those of $FHNC/SOC$ ($APR(LBFIS)$) [41] ([39]) and $AFDMC$ [39], for which the $AV'8$ [23] potential is considered.

The $LOCV$ symmetric nuclear (pure neutron) matter equation of states [19] for the $AV18$ interaction, which are labelled by $MT1$,2, are given in the panel (a(b)) of the figure 3. The $MT1(2)$ results are obtained by including the tensor(spin-orbit)-dependent correlation function in the $^3P_2 -^3 F_2$ channel. Also, the $FHNC/SOC$ calculations of the references [41] (APR) and [39] ($LBFIS$) are plotted for comparison. Note that in the $FHNC/SOC$ computations of $APR(LBFIS)$, the $AV18\,(AV'8)$ [22, 23] is considered. Moreover, the symmetric nuclear (pure neutron) matter $AFDMC$ predictions of the reference [25] ([39]), using the $AV'7(8)$ potential [25] ([23]), are shown by squares in the mentioned figure.

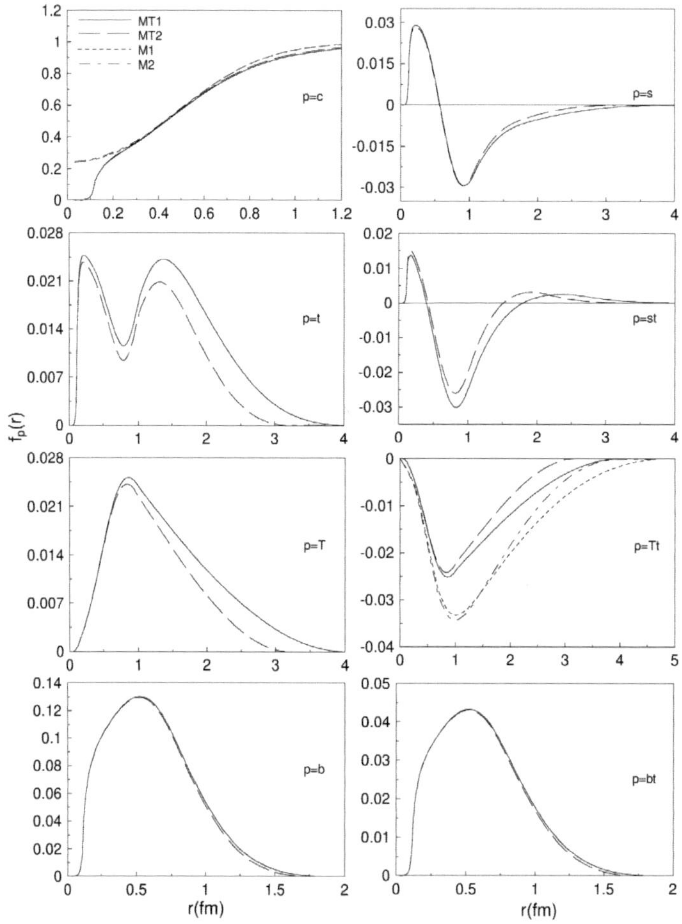

Figure 4. The $LOCV$ nuclear matter correlation functions $f_p(r)$ [19], using the $AV18$ [22] interaction, at $\rho = 0.16\,(0.24) fm^{-3}$, are labelled by $MT1(2)$. The corresponding $FHNC/SOC$ results [36] ($M1, 2$), employing the $AV18$ and the UIX potentials [43, 44], are also shown for comparison. Note that according to the first 8 operators of the $AV18$ interaction [22], $f_p(r)$, for $p = c, s, t, st, T, Tt, b, bt$, are respectively plotted in 8 panels.

As it is demonstrated in the figure 3, the $MT1$ results fairly agree with those of $FHNC/SOC$ [39, 41] and $AFDMC$ [25, 39], especially at low densities. While, the $MT2$ calculations considerably lie below of the other results, particularly at high densities. So, it can be inferred that, including the spin-orbit-dependent correlation function in the $^3P_2 -^3 F_2$ channel, the *nucleonic* matter equation of state becomes more bound. Furthermore, the $LBFIS$ [39] and the $AFDMC$ [25, 39] predictions disagree with those of $MT1, 2$. Note that: (i) The $AV18$ interaction is employed in the $LOCV$ method. Whereas, in the $LBFIS$ and the $AFDMC$ calculations [22,35], the $AV'7, 8$ potentials are included. (ii) The $LOCV$ results are plotted

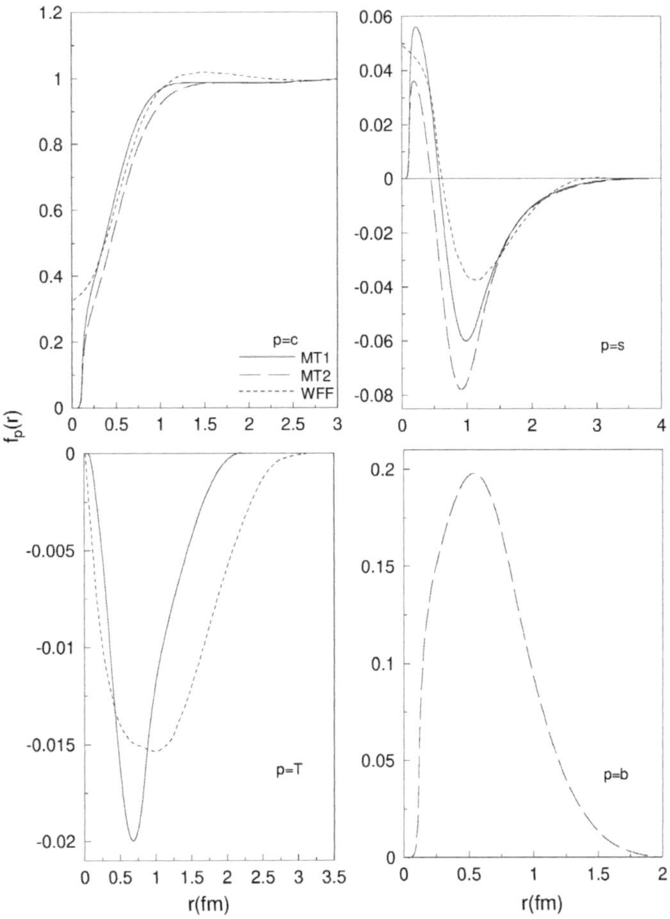

Figure 5. The $LOCV$ neutron matter correlation functions $f_p(r)$ $(MT1,2)$ [19], at $\rho = 0.15fm^{-3}$ in comparison with those $FHNC/SOC$ [35] (WFF). Moreover, the $MT1(2)$ results are calculated, including the tensor-(spin-orbit)-dependent correlation function in the $^3P_2-^3F_2$ channel. Also, the $WFF(LOCV)$ results are given, for the $Av_{14(18)}$ interaction [21] ([22]). Note that $f_p(r)$, for $p = c, s, T, b$, are respectively plotted in 4 panels.

at high densities, i.e. at densities larger than 6 times nuclear saturation density, in order to compare the mentioned calculations with the available $FHNC/SOC$ data. (iii) The presented results for the symmetric nuclear matter, do not show the saturation near the empirical point. Because, the three-body potential is not included in the given calculations.

Now, we display the $LOCV$ *nucleonic* matter operator-dependent correlation functions [19] versus those of $FHNC/SOC$ [35, 36]. In this way, in the figure 4, the $LOCV$ symmetric nuclear matter correlation functions $f_p(r)$, at $\rho = 0.16(0.24)fm^{-3}$ $(MT1(2))$ are plotted in comparison of the $FHNC/SOC$ calculations of the reference [36] $(M1(2))$. Note that: (i)

$f_p(r)$, for $p = c, s, t, st, T, Tt, b, bT$, are respectively given in eight panels. (ii) In the $LOCV$ computations, the $AV18$ interaction is applied. While, in the $FHNC/SOC$ approach, the $AV18$ and the UIX [43, 44] potentials are included. (iii) In the reference [36], the central and the tensor-*isospin*-dependent correlation functions were reported.

Considering the figure 4, one may recognize that the overall behaviour of the $MT1, 2$ results agrees with those of $M1, 2$. Although, the $FHNC/SOC$ results are different from those of $LOCV$. This discrepancies refer to the inclusion of the three-body potential and the many-body terms in the $FHNC/SOC$ calculations. By increasing the density, the $LOCV$ and the $FHNC/SOC$ correlation functions expectedly become short-range. However, the range of the spin-orbit-dependent correlation functions do not change noticeably. Furthermore, the (non-)central correlation functions correctly tend to (zero) one.

In the figure 5, the $LOCV$ pure neutron matter correlation functions $f_p(r)$ [19], using the $AV18$ potential, are presented at $\rho = 0.15fm^{-3}$. In the $LOCV$ formalism, once the tensor-dependent correlation function and once again the spin-orbit-dependent correlation function is employed in the $^3P_2 -^3 F_2$ channel, which are respectively labelled by $MT1$ and $MT2$. The $FHNC/SOC$ correlation functions (WFF) [35], considering the Av_{14} potential [21], are also plotted for comparison. Note that in the reference [35], the correlation functions are obtained by considering the three-body potential $UrbanaVII$ [45]. So, we divide them by the variational parameters β_p [35], in order to obtain the appropriate correlation functions for the two-body interaction Av_{14}. Furthermore, in the mentioned figure, $f_{c,s,T,b}(r_{12})$ are reported. Because, in the neutron matter calculations, the *isospin*-dependent correlation functions cannot be distinguished from those of *isospin*-independent.

As shown in the figure 5, the $LOCV$ results fairly behave like those of WFF. But, the $FHNC/SOC$ spin- and tensor-dependent correlation functions are nearly inconsistent with those of $LOCV$. Because, unlike the $LOCV$ results, the WFF correlation functions are designed for the many-body calculations. Note that the spin-orbit-dependent correlation function, which was not reported in the $FHNC/SOC$ computations, has the shortest range in comparison with the other ones. In addition, at large inter-particle distances, the (non-)central correlation functions expectedly tend to (zero) one.

In the end it is worth mentioning that:

(i) There are some disagreements between the $LOCV$ and the $FHNC/SOC$ predictions. Because, unlike the approximated $LOCV$ approach, in the $FHNC/SOC$ method, the contribution of the many-body terms is included.

(ii) As it was shown in the previous section, the $LOCV$ correlation functions are channel-dependent, i.e. they depend on $JSTT_z$. Furthermore, as it was illustrated in the figure 2, the $ELOCV$ results lie near those of $LOCV$, especially at low densities. This means that, applying the normalization constraint, the contribution of the higher cluster terms than the two-body ones becomes negligible, i.e. the two-body cluster approximation of the $LOCV$ calculations is reliable. As a result, referring to the reference [37], one may expect that the state-independent $FHNC/SOC$ correlation functions could not be as optimal as those of $LOCV$.

(iii) Since, the minimization of the many-body energy with respect to the correlation functions are sophisticated, the $FHNC/SOC$ correlation functions cannot be extracted from the many-body formalism, directly. While, in the $LOCV$ approach, the two-body cluster energy is functionally minimized, subject to the normalization constraint of the radial distribution function, at the two-body cluster approximation. As a result, unlike the $FHNC/SOC$ formalism, the $LOCV$ correlation functions could be extracted from the two-body cluster approximation formalism, directly.

4. Conclusions

In conclusion, we can express that:

(1) The $(E)LOCV$ *Bethe* homework calculations fairly agree with those of $FHNC$ (see figure 2). While, the $(E)LOCV$ results were reported at the two(three)-body cluster approximation. So, one may realize that in the $(E)LOCV$ approach, the contribution of the many-body cluster terms, which is included in the $FHNC$ formalism, can be neglected. Because, in the $(E)LOCV$ computations, we impose the normalization constraint.

(2) The $(E)LOCV$ correlation functions have distinguished features (see figure 1). So, the $FHNC$ calculations with these correlation functions lie close to the $FHNC$ calculations, using the parametrized KT correlation functions (see figure 2).

(3) The $LOCV$ *nucleonic* matter computations, using the sophisticated $AV18$ potential, are fairly consistent with those of $FHNC/SOC$ and $AFDMC$ (see figure 3). Whereas, the $LOCV$ channel-dependent calculations were obtained at the lowest order approximation. It should be note that in the $LOCV$ formalism, the tensor- and spin-orbit-dependent correlation functions were included in the coupled channels. While, as it was noted in the Introduction, the spin-orbit correlations have been recently ignored in the $FHNC/SOC$ calculations.

(4) The behaviour of the $LOCV$ operator-dependent correlation functions is fairly similar to that of $FHNC/SOC$ (see figures 4 to 5).

(5) The many-body effects make the $FHNC/SOC$ correlation functions become different from those of $LOCV$ (see figures 4 to 5). But, as it was noted in the Introduction, one may expect that: (i) The $LOCV$ operator-dependent correlation functions, can improve the convergence of the $FHNC/SOC$ calculations. (ii) Including the $LOCV$ or the $FHNC/SOC$ operator-dependent correlation functions, the $FHNC/SOC$ energy does not vary remarkably.

Acknowledgments

A. Tafrihi would like to thank the Isfahan University of Technology for his support. M. Modarres would like to thank the Research Council of University of Tehran and Institute for Research and Planning in Higher Education for the grants provided for him.

References

[1] V. R. Pandharipande, Nucl. Phys. A **174** (1971) 641;
[2] V. R. Pandharipande, Nucl. Phys. A **178** (1971) 123;
[3] V. R. Pandharipande and H. A. Bethe, Phys. Rev. C **7** (1973) 1312.
[4] J.C. Owen, R.F. Bishop, J.M. Irvine, Phys. Lett. B **59** (1975) 1.
[5] J.C. Owen, R.F. Bishop, J.M. Irvine, Ann. Phys. **102** (1976) 170.
[6] J.C. Owen, R.F. Bishop, J.M. Irvine, Phys. Lett. B **61** (1976) 147.
[7] J.C. Owen, R.F. Bishop and J.M. Irvine, Phys. Lett.B, **66** (1977) 25.
[8] M. Modarres and J.M. Irvine, J. Phys. G: Nucl. Phys. **5** (1979) 7.
[9] M. Modarres, Europhys. Lett. **3** (1987) 1083.
[10] M. Modarres, J. Phys. G: Nucl. Phys. **21** (1995) 351.
[11] G.H. Bordbar, M. Modarres, Phys. Rev. C **57** (1998) 714.
[12] M. Modarres and G.H. Bordbar, Phys. Rev. C **58** (1998) 2781.
[13] M. Modarres and H.R. Moshfegh, Phys. Rev. C **62** (2000) 044308.
[14] M. Modarres, A. Rajabi and H.R. Moshfegh, Phys. Rev. C **76** (2007) 064311.
[15] M. Modarres, A. Rajabi, H.R. Moshfegh, Nucl. Phys. A **808** (2008) 60.
[16] M. Modarres, M. Pourmirjafari, H.R. Moshfegh, Nucl. Phys. A **819** (2009) 27.
[17] M. Modarres, A. Tafrihi, A. Hatami, Nucl. Phys. A **879** (2012) 1.
[18] M. Modarres, A. Tafrihi, Nucl. Phys. A **916** (2013) 126.
[19] M. Modarres, A. Tafrihi, Nucl. Phys. A **941** (2015) 212.
[20] R.V. Reid, Ann. Phys. **50** (1968) 411.
[21] R.B. Wiringa, R.A. Smith, T.L. Ainsworth, Phys. Rev. C **29** (1984) 1207.

[22] R.B. Wiringa, V.G.J. Stoks, R. Schiavilla, Phys. Rev. C **51** (1995) 38.
[23] B.S. Pudliner, V.R. Pandharipande, J. Carlson, S.C. Pieper, R.B. Wiringa, Phys. Rev. C **56** (1997) 1720.
[24] R.B. Wiringa, S.C. Pieper, Phys. Rev. Lett. **89** (2002) 182501.
[25] S. Gandolfi, A. Lovato, J. Carlson, and Kevin E. Schmidt, Phys. Rev. C **90** (2014) 061306(R).
[26] F. Iwamoto and M. Yamada: Prog. Theor. Phys. **17** (1957) 543.
[27] J.W. Clark, Prog. Part. Nucl. Phys. **2** (1979) 89.
[28] M.E. Grypeos and E. Mavrommatis, Lett. Nuovo Cim. **5** (1972) 369.
[29] E. Krotscheck and K. Takahashi, Phys.Lett.B, **61** (1976) 393.
[30] S. Fantoni and S. Rosati, Lett.Nuovo.Cim., **16** (1976) 531.
[31] J.G. Zabolitzky, Phys.Rev.A, **16** (1977) 1258.
[32] S. Fantoni and K.E. Schmidt, Nucl.Phys.A, **690** (2001) 456.
[33] N. Bassan, S. Fantoni and K.E. Schmidt, Phys.Rev.C, **84** (2011) 035807.
[34] V.R. Pandharipande, R.B. Wiringa, Rev. Mod. Phys. **51** (1979) 821.
[35] R.B. Wiringa, V. Fiks, A. Fabrocini, Phys. Rev. C **38** (1988) 1010.
[36] A. Mukherjee, Phys. Rev. C **79** (2009) 045811.
[37] A. Akmal, V. Pandharipande, Phys. Rev. C **56** (1997) 2261.
[38] A. Lovato, PhD thesis at SISSA-ISAS, Trieste, Italy (2012).
[39] Alessandro Lovato, Omar Benhar, Stefano Fantoni, Alexey Yu. Illarionov, Kevin E. Schmidt, J. Phys.: Conf. Ser. **336** (2011) 012016.
[40] M. Baldo, A. Polls, A. Rios, H.-J. Schulze, and I. Vidaña, Phys. Rev. C **86** (2012) 064001.
[41] A. Akmal, V.R. Pandharipande, D.G. Ravenhall, Phys. Rev. C **58** (1998) 1804.
[42] D. Ceperley, G. V. Chester, M. H. Kalos, Phys. Rev. B **16** (1977) 3081.
[43] J. Calson, V.R. Pandharipande, R.B. Wiringa, Nucl. Phys. A **401** (1983) 59.
[44] B.S. Pudliner, V.R. Pandharipande, J. Carlson, R.B. Wiringa, Phys. Rev. Lett. **74** (1995) 4396.
[45] R. Schiavilla, V.R. Pandharipande, R.B. Wiringa, Nucl. Phys. A **449** (1986) 219.

Variational calculations with explicit energy functionals for fermion systems at zero temperature

M Takano[1,2], T Suzuki[2] and N Sakumichi[3]

[1]Research Institute for Science and Engineering, Waseda University
3-4-1 Okubo, Shinjuku-ku, Tokyo 169-8555 Japan.
[2]Department of Physics and Applied Physics, Waseda University
3-4-1 Okubo, Shinjuku-ku, Tokyo 169-8555 Japan.
[3]Center for Soft Matter Physics, Ochanomizu University
2-2-1 Otsuka, Bunkyo-ku, Tokyo 112-8610 Japan.

E-mail: takanom@waseda.jp

Abstract. The variational method with explicit energy functionals (EEFs) is applied to infinite neutron matter. Starting from the Argonne v8' two-body potential and the repulsive part of the Urbana IX three-body potential, the energy per neutron is expressed explicitly with the spin-dependent central, tensor, and spin-orbit distribution functions. This EEF is constructed as an extension of the previously proposed one for the Argonne v6' potential, including central and tensor forces. The Euler-Lagrange equations derived from this EEF are solved numerically. The obtained fully minimized energy with this EEF is in good agreement with that obtained from the auxiliary field diffusion Monte Carlo calculation.

1. Introduction

The variational method with explicit energy functionals (EEFs) was proposed in Ref. [1] for calculating energy per particle of spin-1/2 strongly correlated fermion systems, such as nuclear matter and liquid ^{3}He at zero temperature. In this method, the energy per particle of a system in which two particles interact through a two-body central force is expressed explicitly with spin-dependent radial distribution functions, which are regarded as variational functions. Then, the Euler-Lagrange equations for these radial distribution functions derived from the EEF are solved numerically, and the fully minimized energy per particle is obtained.

Here we compare this EEF variational method with the Fermi Hypernetted Chain (FHNC) variational method[2]. In the FHNC method, the expectation value of the Hamiltonian per particle $\langle H\rangle/N$ with the Jastrow wave function is calculated approximately based on the cluster variational method originating from Ref. [3]. Then, the $\langle H\rangle/N$ is minimized with respect to the correlation function included in the Jastrow wave function. Namely, in the case of the FHNC method, it is clear which cluster terms are included in $\langle H\rangle/N$. However, the correlation function is usually parameterized for nuclear matter and the minimization of $\langle H\rangle/N$ is performed with respect to those parameters. This implies that the minimization of $\langle H\rangle/N$ may be insufficient in the FHNC method. In contrast, the fully-minimized energy per particle is obtained in the variational method with the EEF, which is an advantageous point of this method. Further comparison with the FHNC method is given in Refs.[1, 4], and is also shown below.

The energy per particle of liquid ^{3}He calculated with this variational method is in fairly good agreement with the experimental data [1]. Moreover, a refinement of the EEF with respect to a part of the three-body cluster terms was reported in Ref. [5]. To treat nuclear matter, in Ref. [6], the EEF was extended to include the tensor and spin-orbit forces. Because this extension was observed to be insufficient, further refinements of the EEF were made with respect to the kinetic energy caused by the tensor correlations, as reported in Ref. [4]. Furthermore, a state-independent three-body nuclear force was considered, and the EEF of neutron matter at zero temperature was extended to finite temperatures in Ref. [4].

In this paper, the variational method with the EEF is applied to infinite neutron matter including the two-body spin-orbit force.

2. Explicit energy functional of neutron matter with the spin-orbit force

In this section, we report on an improvement of the EEF for neutron matter so as to take into account the spin-orbit nuclear force. In a previous study [4], we proposed an EEF with the Argonne v6' (AV6) two-body potential [7] and the repulsive part of the Urbana IX (UIX) three-body potential [8] for neutron matter at zero temperature.

In order to extend the EEF so as to treat the spin-orbit force, we newly introduce the spin-orbit distribution functions $F_{\rm SO}(r)$ in addition to the radial distribution functions $F_s(r)$ and tensor distribution function $F_{\rm T}(r)$ employed in the previous EEF. These functions are defined as follows:

$$F_s(r_{12}) = \Omega^2 \sum_{\rm spin} \int \Psi^\dagger(x_1, x_2, ..., x_N) P_{s12} \Psi(x_1, x_2, ..., x_N) d\mathbf{r}_3 d\mathbf{r}_4 ... d\mathbf{r}_N, \quad (1)$$

$$F_{\rm T}(r_{12}) = \Omega^2 \sum_{\rm spin} \int \Psi^\dagger(x_1, x_2, ..., x_N) S_{\rm T12} \Psi(x_1, x_2, ..., x_N) d\mathbf{r}_3 d\mathbf{r}_4 ... d\mathbf{r}_N, \quad (2)$$

$$F_{\rm SO}(r_{12}) = \Omega^2 \sum_{\rm spin} \int \Psi^\dagger(x_1, x_2, ..., x_N) (\mathbf{L}_{12} \cdot \mathbf{s}) \Psi(x_1, x_2, ..., x_N) d\mathbf{r}_3 d\mathbf{r}_4 ... d\mathbf{r}_N. \quad (3)$$

Here, Ω represents the volume of neutron matter, and the summations on the right-hand sides of these equations are conducted over the spin coordinates of all the neutrons. In addition, Ψ is the wave function of neutron matter, and x_i represents the spatial coordinate $\mathbf{r}_i$ and the spin coordinate of the i-th neutron. Furthermore, P_{sij} is the spin-projection operator projecting the (i, j) neutron pair state onto the spin triplet ($s = 1$) or spin singlet ($s = 0$) state, $S_{{\rm T}ij}$ is the tensor operator, and $\mathbf{L}_{ij}$ is the relative orbital angular momentum operator.

For convenience, we also introduce the following auxiliary functions, i.e., the intrinsically central distribution functions $F_{{\rm C}s}(r)$, the dressed tensor correlation function $g_{\rm T}(r)$, and the dressed spin-orbit correlation function $g_{\rm SO}(r)$, which are defined as the solutions to the following equations[6]:

$$F_s(r) = F_{{\rm C}s}(r) + 8s \left[g_{\rm T}(r)\right]^2 F_{{\rm F}s}(r) + \frac{2}{3} s \left[g_{\rm SO}(r)\right]^2 F_{{\rm qF}s}(r), \quad (4)$$

$$F_{\rm T}(r) = 16 \left[\sqrt{F_{\rm C1}(r) F_{\rm F1}(r)} g_{\rm T}(r) - \left[g_{\rm T}(r)\right]^2 F_{\rm F1}(r)\right] - \frac{2}{3} \left[g_{\rm SO}(r)\right]^2 F_{\rm qF1}(r), \quad (5)$$

$$F_{\rm SO}(r) = -24 \left[g_{\rm T}(r)\right]^2 F_{\rm F1}(r) + \frac{4}{3} \left[\sqrt{\frac{F_{\rm C1}(r)}{F_{\rm F1}(r)}} g_{\rm T}(r) - \frac{\left[g_{\rm T}(r)\right]^2}{4} - g_{\rm T}(r) g_{\rm SO}(r)\right] F_{\rm qF1}(r). \quad (6)$$

Here, $F_{{\rm F}s}(r)$ ($s = 0, 1$) are the spin-dependent radial distribution functions for the Fermi gas, and $F_{{\rm qF}s}(r)$ are given by

$$F_{{\rm qF}s}(r) = \frac{2s+1}{4} \left[\frac{(k_{\rm F} r^2)}{5} - (-1)^s 9 j_2(k_{\rm F} r) \frac{j_1(k_{\rm F} r)}{k_{\rm F} r}\right]. \quad (7)$$

Note that without the tensor and spin-orbit forces, $F_{Cs}(r)$ reduce to $F_s(r)$, whereas $g_T(r)$ and $g_{SO}(r)$ vanish.

To construct the EEF, the following structure functions are also necessary:

$$S_{c1}(k) = \frac{1}{N}\left\langle\left|\sum_{i=1}^{N}\exp(i\mathbf{k}\cdot\mathbf{r}_i)\right|^2\right\rangle = 1 + S_1(k) + S_0(k) \geq 0, \tag{8}$$

$$S_{cT1}(k) = \frac{1}{Nk^2}\left\langle\left|\sum_{i=1}^{N}(\sigma_i\cdot\mathbf{k})\exp(i\mathbf{k}\cdot\mathbf{r}_i)\right|^2\right\rangle = 1 + \frac{1}{3}S_1(k) - S_0(k) - \frac{S_T(k)}{3} \geq 0, \tag{9}$$

$$S_{cT2}(k) = \frac{1}{2Nk^2}\left\langle\left|\sum_{i=1}^{N}(\sigma_i\times\mathbf{k})\exp(i\mathbf{k}\cdot\mathbf{r}_i)\right|^2\right\rangle = 1 + \frac{1}{3}S_1(k) - S_0(k) + \frac{S_T(k)}{6} \geq 0. \tag{10}$$

Here, $S_s(k)$ and $S_T(k)$ are defined as follows:

$$S_s(k) = 4\pi\rho\int_0^\infty [F_s(r) - F_s(\infty)]\, j_0(kr) r^2 dr, \tag{11}$$

$$S_T(k) = 4\pi\rho\int_0^\infty F_T(r) j_2(kr) r^2 dr, \tag{12}$$

with ρ representing the neutron number density. Furthermore, corresponding to the introduction of $F_{SO}(r)$, we also define

$$S_{SO}(k) = 4\pi\rho\int_0^\infty F_{SO}(r)\frac{j_1(kr)}{k_F r} r^2 dr. \tag{13}$$

Here, k_F is the Fermi wave number.

Using these functions, we propose an improved EEF for neutron matter with the spin-orbit force as follows:

$$\frac{E_2}{N} = \frac{3}{5}E_F + \frac{E_V}{N} + \frac{E_{TF}}{N} + \frac{E_{TS}}{N} + \frac{E_{TN}}{N}. \tag{14}$$

Here, the first term on the right-hand side of Eq. (14) represents the one-body kinetic energy, and E_F corresponds to the Fermi energy. The second term E_V/N represents the potential energy given by

$$\frac{E_V}{N} = 2\pi\rho\int_0^\infty \left[\sum_{s=0}^{1} F_s(r)V_{Cs}(r) + F_T(r)V_T(r) + F_{SO}(r)V_{SO}(r)\right] r^2 dr, \tag{15}$$

with $V_{Cs}(r)$, $V_T(r)$, and $V_{SO}(r)$ being the two-body central, tensor, and spin-orbit potentials, respectively. The remaining terms E_{TF}/N, E_{TS}/N, and E_{TN}/N on the right-hand side of Eq. (14) represent the kinetic energy caused by the correlations among neutrons expressed as

$$\begin{aligned}\frac{E_{TF}}{N} &= \frac{\pi\hbar\rho}{2m}\int_0^\infty \sum_{s=0}^{1} F_{Cs}(r)\left[\frac{1}{F_{Cs}(r)}\frac{dF_{Cs}(r)}{dr} - \frac{1}{F_{Fs}(r)}\frac{dF_{Fs}(r)}{dr}\right]^2 r^2 dr \\ &+ \frac{2\pi\hbar\rho}{m}\int_0^\infty \left[8\left\{\left[\frac{dg_T(r)}{dr}\right]^2 + \frac{6}{r^2}[g_T(r)]^2\right\}F_{F1}(r) + \frac{2}{3}\left[\frac{dg_{SO}(r)}{dr}\right]^2 F_{qF1}(r)\right] r^2 dr,\end{aligned} \tag{16}$$

$$\begin{aligned}\frac{E_{TS}}{N} &= -\frac{\hbar}{16\pi m\rho}\int_0^\infty \frac{[S_{c1}(k) - S_{ex}(k)]\left\{[S_{c1}(k) - S_{cF}(k)]^2 + \frac{15}{2}[S_{SO}(k)]^2\right\}}{S_{c1}(k)/S_{cF}(k)} \\ &- \frac{\hbar}{16\pi m\rho}\int_0^\infty \sum_{n=1}^{2} n\frac{[S_{cTn}(k) - S_{ex}(k)]\left\{[S_{cTn}(k) - S_{cF}(k)]^2 + \frac{15}{4}(n-1)[S_{SO}(k)]^2\right\}}{S_{cTn}(k)/S_{cF}(k)} k^4 dk,\end{aligned} \tag{17}$$

with $S_{\rm ex}(k) = 2S_{\rm cF}(k) - 3$, and

$$\frac{E_{\rm TN}}{N} = \frac{\hbar\rho^2}{2m}\sum_{n=1}^{4} c_n a_n b_n, \tag{18}$$

$$a_1 = \rho\int\left\{\left[\frac{dG_{\rm C}(r)}{dr}\right]^2 + 2G_{\rm T}(r)\right\}d\mathbf{r},\ b_1 = \rho\int\left\{[G_{\rm C}(r)]^2 + 8\,[g_{\rm T}(r)]^2\right\}d\mathbf{r},$$
$$a_2 = b_2 = k_{\rm F}\rho\int g_{\rm T}(r)g_{\rm SO}(r)d\mathbf{r},\ a_3 = \rho\int\left\{\frac{1}{20}\left[\frac{dG_{\rm C}(r)}{dr}\right]^2 + \frac{2}{5}G_{\rm T}(r)\right\}d\mathbf{r},$$
$$b_3 = \rho\int[g_{\rm SO}(r)]^2(k_{\rm F}r)^2 d\mathbf{r},\ a_4 = \rho\int\left\{\left[\frac{dg_{\rm SO}(r)}{dr}\right]^2 + \frac{6}{r^2}[g_{\rm SO}(r)]^2\right\}(k_{\rm F}r)^2 d\mathbf{r},$$
$$b_4 = \rho\int\left\{\frac{1}{20}[G_{\rm C}(r)]^2 + \frac{2}{5}[g_{\rm T}(r)]^2 + \frac{17}{4800}[g_{\rm SO}(r)]^2(k_{\rm F}r)^2\right\}d\mathbf{r} \tag{19}$$

with

$$G_{\rm C}(r) = \frac{1}{3}F_{\rm C1}(r) - F_{\rm C0}(r) - \frac{1}{3}F_{\rm F1}(r) + F_{\rm F0}(r),\ G_{\rm T}(r) = \left[\frac{dg_{\rm T}(r)}{dr}\right]^2 + \frac{6}{r^2}[g_{\rm T}(r)]^2, \tag{20}$$

and $(c_1, c_2, c_3, c_4) = (3/4, -1/60, 1, 1)$.

The EEF given in Eq. (14) is constructed with the following procedure. The first and the second terms are the exact expressions regardless of the explicit functional form of the wave function $\Psi(x_1, x_2, ..., x_N)$. In order to construct the rest of the kinetic-energy terms, we temporarily assume the following Jastrow wave function

$$\Psi = {\rm Sym}\left\{\prod_{i<j}\left[\sum_{s=0}^{1} f_{{\rm C}s}(r_{ij})P_{sij} + f_{\rm T}(r_{ij})S_{{\rm T}ij} + f_{\rm SO}(r_{ij})\left(\mathbf{L}_{ij}\cdot\mathbf{s}\right)\right]\right\}\Phi_{\rm F}. \tag{21}$$

Here, $\Phi_{\rm F}$ is the wave function of the Fermi gas. The meaning of temporarily will be discussed below. Then, $E_{\rm TF}/N$ is constructed so that E_2/N properly includes the two-body cluster kinetic energy terms in $\langle H\rangle/N$. It is noted that $E_{\rm TF}/N$ also includes a part of the three-body cluster and higher-order cluster kinetic energy terms in $\langle H\rangle/N$, because it is expressed with $F_s(r)$, $F_{\rm T}(r)$, and $F_{\rm SO}(r)$ rather than $f_{{\rm C}s}(r)$, $f_{\rm T}(r)$, and $f_{\rm SO}(r)$. In fact, the three-body cluster kinetic energy direct terms that are the lowest order in $h_{{\rm C}s}(r) = f_{{\rm C}s}(r) - 1$, $f_{\rm T}(r)$, and $f_{\rm SO}(r)$, which are denoted by $E_{\rm 3d}/N$, are included partially in $E_{\rm TF}/N$. The rest of $E_{\rm 3d}/N$ is expressed as the form of $E_{\rm TS}/N$ with the denominators $S_{\rm c1}(k)$ and $S_{{\rm cT}n}(k)$ being replaced by unity. Then, we introduce the denominators into $E_{\rm TS}/N$ for the following two reasons. (i) $E_{\rm TS}/N$ without the denominators is a harmful term which goes to negative infinity when the E_2/N without these denominators is minimized with respect to $F_s(r)$, $F_{\rm T}(r)$ and $F_{\rm SO}(r)$: The harmful term is converted into a harmless term by the introduction of the denominators. (ii) $S_{\rm c1}(k)$ and $S_{{\rm cT}n}(k)$ ($n = 1$, 2) are non-negative functions as shown in Eqs. (8)-(10)[9], and the introduction of the denominators in the expression of $E_{\rm TS}/N$ guarantees the non-negativeness of those structure functions. It is noted that in $E_{\rm 3d}/N$ including the spin-orbit operator $(\mathbf{L}_{ij}\cdot\mathbf{s})$, we consider for simplicity only the terms in which $\mathbf{L}_{ij}$ operates on $\Phi_{\rm F}$. In addition, as reported in Ref. [4], the two-particle exchange three-body cluster terms that are the lowest order in $h_{{\rm C}s}(r)$ and $f_{\rm T}(r)$ are included in $E_{\rm TF}/N$ and $E_{\rm TS}/N$. The last term on the right-hand side of Eq. (14), $E_{\rm TN}/N$, represents the three-body cluster direct nodal diagrams that are the lowest order in $h_{{\rm C}s}(r)$, $f_{\rm T}(r)$, and $f_{\rm SO}(r)$.

In addition to the two-body potential given above, we consider the state-independent three-body potential $V_3(\mathbf{r}_i, \mathbf{r}_j, \mathbf{r}_k)$ by introducing the corresponding three-body distribution function $F_3(\mathbf{r}_i, \mathbf{r}_j, \mathbf{r}_k)$. Then, the total energy per neutron E/N is expressed as follows:

$$\frac{E}{N} = \frac{E_2}{N} + \frac{\rho^2}{6} \int F_3(\mathbf{r}_1, \mathbf{r}_2, \mathbf{r}_3) V_3(\mathbf{r}_1, \mathbf{r}_2, \mathbf{r}_3) d\mathbf{r}_{12} d\mathbf{r}_{23}. \tag{22}$$

Here, as in Ref. [4], we employ the following extended Kirkwood's assumption:

$$F_3(\mathbf{r}_1, \mathbf{r}_2, \mathbf{r}_3) = F(r_{12})F(r_{23})F(r_{31}) \frac{F_{3\mathrm{F}}(\mathbf{r}_1, \mathbf{r}_2, \mathbf{r}_3)}{F_{\mathrm{F}}(r_{12})F_{\mathrm{F}}(r_{23})F_{\mathrm{F}}(r_{31})}, \tag{23}$$

where $F(r) = F_1(r) + F_0(r)$, and $F_{3\mathrm{F}}(\mathbf{r}_1, \mathbf{r}_2, \mathbf{r}_3)$ is the three-body distribution function $F_3(\mathbf{r}_1, \mathbf{r}_2, \mathbf{r}_3)$ for the Fermi gas.

It is worth noting that a variational method which treats more cluster terms in $\langle H \rangle / N$ with the Jastrow wave function is not necessarily a better variational method, because the Jastrow wave function is an approximate one which cannot express the exact ground-state wave function of neutron matter: As more cluster terms are included, $\langle H \rangle / N$ approaches the exact expectation value with the Jastrow wave function, which is just the upper bound of the true ground-state energy, even if the minimization is performed sufficiently. Furthermore, the Jastrow wave function Eq. (21) is not appropriate for representing the spin dependence of the correlation among neutrons, as pointed out in Ref.[1]. This shortcoming is found in the three-body cluster and higher-order cluster terms: It is not necessarily a better approximation to taking into account more cluster terms sophisticatedly. In the present EEF method, therefore, the energy per neutron E/N is directly connected to the two-body distribution functions $F_s(r)$, $F_{\mathrm{T}}(r)$, and $F_{\mathrm{SO}}(r)$ (or practically, $F_{\mathrm{C}s}(r)$, $g_{\mathrm{T}}(r)$, and $g_{\mathrm{SO}}(r)$). We do not persist in treating the three-body cluster and higher-order cluster kinetic energy terms rigorously in the present EEF variational method: A part of the three-body cluster kinetic-energy terms in $\langle H \rangle / N$ are used just as a reference for the construction of the corresponding part of EEF. This is the reason why we temporarily assume the Jastrow wave function.

3. Results and Future Perspectives

Using the new EEF shown above, we derive the Euler-Lagrange (EL) equations for $F_{\mathrm{C}s}(r)$, $g_{\mathrm{T}}(r)$, and $g_{\mathrm{SO}}(r)$ and solve them numerically for the Argonne v8' two-body potential [10] and UIX three-body potential to obtain fully minimized E/N. The EL equations are the coupled integrodifferencial equations, and are solved in the iterative way: Detailed procedure of the numerical calculations is similar to those given in Ref.[6]. The result is presented in Figure 1 together with the result by the auxiliary field diffusion Monte Carlo (AFDMC) calculation [11]. The present result is in good agreement with that obtained from the AFDMC calculation. We note that in the AFDMC calculation, the 2π-exchange part of the UIX potential is also included. Therefore, this agreement implies that the contribution from the 2π-exchange part of the UIX potential is negligibly small in E/N of neutron matter. We note that the numerical calculations for the EEF variational method is possibly much less the CPU demanding than the AFDMC calculation, which is an advantageous point for the present variational method.

In future work, the evaluation of the three-body cluster terms excluded in the present energy expression is important. Furthermore, it is interesting to include the 2π-exchange part of the UIX potential to directly evaluate its contribution. Another important future problem would be to extend the present EEF to symmetric nuclear matter. In addition, it is interesting to apply the present EEF variational method to low dimensional systems, such as a two-dimensional (2D) ^{3}He system. In the strictly 2D ^{3}He system, the Monte Carlo calculations show no negative-energy point at finite number density, which implies that the 2D ^{3}He system is a gas state at zero

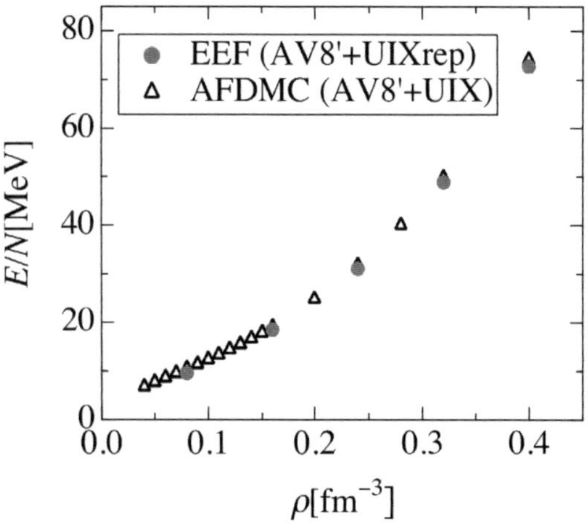

Figure 1. Energy per neutron of neutron matter with the AV8' two-body potential and the repulsive part of the UIX three-body potential as a function of the number density ρ. The result obtained from the AFDMC calculation with the full UIX potential is also shown.

temperature[12, 13]. Here, it is pointed out that the Fermi statistics play a crucial role. On the other hand, recent experiments observed a self-bound liquid state in the 2D ^{3}He system, by using ^{3}He atoms absorbed on graphite [14]. Under these paradoxical circumstances, we extended our simplest EEF given in Ref. [1] to strict 2D ^{3}He system, and obtained a negative-energy minimum point as a preliminary numerical result. In order to examine whether this minimum point is an artifact stemming from the insufficiency with respect to the Fermi statistics, the EEF is now being improved with respect to the three-body cluster exchange terms, as reported in Ref. [5] for 3D liquid ^{3}He. The results will be reported elsewhere in the near future.

Acknowledgments

The numerical computations in this work were carried out on SR16000 at YITP in Kyoto University and on SR16000 at the University of Tokyo. This work is supported by JSPS (No. 25400275) and the Waseda University Grant for Special Research Projects (Project number: 2015K-200). N.S. is supported by a Grant-in-Aid for JSPS Fellows (Grant No. 250588) and by JSPS KAKENHI Grant Number 15K17725.

References

[1] Takano M and Yamada M 1994 *Prog. Theor. Phys.* **91** 1149
[2] Clark J W, 1979 *Prog. Part. Nucl. Phys.* **2** 89
[3] Iwamoto F and Yamada M, 1957 *Prog. Theor. Phys.* **17** 543.
[4] Takano M, Kato K and Yamada M 2014 *J. Phys.: Conf. Ser.* **529** 012025
[5] Takano M, Kaneko T and Yamada M 1997 *Prog. Theor. Phys.* **97** 569
[6] Takano M and Yamada M 1998 *Prog. Theor. Phys.* **100** 745
[7] Wiringa R B and Pieper S C 2002 *Phys. Rev. Lett.* **89** 182501
[8] Carlson J, Pandharipande V R, Wiringa R B, 1983 *Nucl. Phys.* A **401** 59
[9] Takano M and Yamada M 1992 *Prog. Theor. Phys.* **88** 1131
[10] Pudliner B S, Pandharipande V R, Carlson J, Pieper S C and Wiringa R B 1997 *Phys. Rev.* C **56** 1720
[11] Gandorfi S, Carlson J, Reddy S, Steiner A W and Wiringa R B 2014 *Eur. Phys. J.* **A50** 10
[12] Grau V, Boronat J and Casulleras J 2002 *Phys. Rev. Lett.* **89** 045301
[13] Ruggeri M, Moroni S and Boninsegni M 2013 *Phys. Rev. Lett.* **111** 045303
[14] Sato D, Naruse K, Matsui T and Fukuyama H 2012 *Phys. Rev. Lett.* **109** 235306